AF677537

Competitive Quality and Innovation

Pierre Maillard

iSTE

WILEY

First published 2015 in Great Britain and the United States by ISTE Ltd and John Wiley & Sons, Inc.

Apart from any fair dealing for the purposes of research or private study, or criticism or review, as permitted under the Copyright, Designs and Patents Act 1988, this publication may only be reproduced, stored or transmitted, in any form or by any means, with the prior permission in writing of the publishers, or in the case of reprographic reproduction in accordance with the terms and licenses issued by the CLA. Enquiries concerning reproduction outside these terms should be sent to the publishers at the undermentioned address:

ISTE Ltd
27-37 St George's Road
London SW19 4EU
UK

www.iste.co.uk

John Wiley & Sons, Inc.
111 River Street
Hoboken, NJ 07030
USA

www.wiley.com

© ISTE Ltd 2015
The rights of Pierre Maillard to be identified as the author of this work have been asserted by him in accordance with the Copyright, Designs and Patents Act 1988.

Library of Congress Control Number: 2015937458

British Library Cataloguing-in-Publication Data
A CIP record for this book is available from the British Library
ISBN 978-1-84821-820-8

Contents

Preface xiii

Introduction xvii

Chapter 1. The Innovation Concept 1

1.1. The characteristics of an innovation process 1
1.2. Review of basic conceptual definitions in "general systemics" 13
1.2.1. The concept of a process 14
1.2.2. The concept of a processor 14
1.2.3. The concept of functionality 14
1.2.4. The concept of a function 14
1.2.5. The concept of configuration 15
1.2.6. The managerial architecture of a processor 15
1.3. Evaluation criteria for the success of an innovation 17
1.4. Drivers of socioeconomic exchange for an innovation process 18
1.5. Clarifications on certain actions in an innovation process 23
1.5.1. Managing the configuration of an innovation process 23
1.5.2. Creative research 25

1.5.3. Studies on the impact of a new innovation 29
1.5.4. Feasibility studies in the innovation process 32
1.5.5. The decision-making stages of the innovation process 36
1.6. Classification of innovation processes 37
1.6.1. Three types of company 37
1.6.2. Types of innovations 39
1.6.3. Correlations between types of innovating companies and types of innovation 39
1.6.4. The specificities of type I innovations 40
1.6.5. Specificities of type II innovations 44
1.6.6. Specificities of type II and IV innovations 49
1.7. Conclusion 52

CHAPTER 2. COMPETITIVE QUALITY OF AN INNOVATION 53

2.1. Introduction 53
2.2. The concepts which characterize competitive quality 55
2.3. The use of competitive quality in an innovation process 67
2.4. A model of the competitive quality process in an innovation process 69
2.5. Conclusion 73

CHAPTER 3. COMPETITIVE QUALITY TACTICS 75

3.1. Introduction 75
3.2. The objective of a competitive quality tactics 76
3.3. Planning competitive quality tactics in the innovation process 81
3.4. Methods for designing competitive quality tactics 88
3.5. The performance of competitive quality tactics 91

3.6. Managing the design of competitive quality tactics 91
3.7. Conclusion. 94

CHAPTER 4. THE MARKETING QUALITY OF AN INNOVATION 97

4.1. Introduction. 97
4.2. Marketing quality in an innovation process 99
4.3. Planning marketing quality within the innovation process. 103
4.3.1. Releases of competitive quality fuel 103
4.3.2. The emergence of stakeholders' competitive quality perceptions 106
4.3.3. Compensatory resources and contributions to the company's image and reputation 107
4.4. Defining marketing quality 108
4.4.1. Marketing quality in releasing competitive quality fuel. 110
4.4.2. Marketing quality in the emergence of competitive quality judgments 111
4.4.3. Marketing quality in the production of compensatory resources 113
4.4.4. Marketing quality in contributing to the company's image and reputation 114
4.5. Marketing quality performance. 115
4.6. Managing the production of marketing quality. 117
4.7. Conclusion. 120

CHAPTER 5. THE "QUALITY" FUEL OF AN INNOVATION 123

5.1. Introduction. 123
5.2. An innovation process' competitive quality fuel. 124
5.3. Planning the design of competitive quality fuel in the innovation process. 127

5.4. Ways of defining competitive quality fuel 129
5.5. Performance of the design of competitive quality fuel 136
5.6. Managing the design of competitive quality fuel 138
5.7. Conclusion 141

CHAPTER 6. THE EMITTERS OF QUALITY FUEL 143

6.1. Introduction 143
6.2. Competitive quality fuel emitters 144
6.3. Defining a competitive quality fuel emitter 146
6.4. Planning the design of competitive quality fuel emitters 151
6.5. The performance of the design of competitive quality fuel emitters 154
6.6. Managing the design of quality fuel emitters 155
6.7. Conclusion 158

CHAPTER 7. QUALIFICATION OF EMITTERS 159

7.1. Introduction 159
7.2. The qualification of competitive quality fuel emitters 160
7.3. Method for qualifying competitive quality emitters 161
7.4. The performance of this qualification 164
7.5. Managing the qualification of competitive quality fuel emitters 165
7.6. Conclusion 168

CHAPTER 8. VALIDATION OF MARKETING QUALITY 169

8.1. Introduction 169
8.2. Validation of the competitive quality marketing of an innovation process 170
8.3. Methods for validating competitive marketing quality in an innovation process 171

8.4. Planning the validation of competitive marketing quality 172
8.5. Managing competitive marketing quality validation 173
8.6. Conclusion 175

CHAPTER 9. SUPERVISION OF EMITTERS 177

9.1. Introduction 177
9.2. The objective of supervision 178
9.3. Methods of supervision 179
9.4. Planning the supervision of competitive quality fuel emitters 180
9.5. The performance of the supervision of CQF emitters 182
9.6. Managing the supervision of competitive quality fuel emitters 183
9.7. Conclusion 184

CHAPTER 10. MONITORING PERCEIVED QUALITY 187

10.1. Introduction 187
10.2. The objective of this monitoring 188
10.3. Methods of monitoring perceived quality 189
10.4. Planning the monitoring of perceived quality 190
10.5. Managing the monitoring of perceived quality 192
10.6. Conclusion 194

CHAPTER 11. ENSURING BENEFITS 197

11.1. Introduction 197
11.2. The objective of ensuring benefits 198
11.3. Methods for ensuring benefits 199
11.4. Planning methods for ensuring benefits 200
11.5. The management of ensuring benefits 202
11.6. Conclusion 203

CHAPTER 12. THE ROLE OF THE QUALITY DEPARTMENT 205

12.1. Introduction 205
12.2. Positioning a quality department within an innovation process 206
12.3. The specificities of the quality specialist's objectives 211
12.4. The objectives of the quality specialist in creative research 214
12.4.1. The quality specialist's interventions in the emergence of the idea 216
12.4.2. The interventions of the quality specialist in the creation of experimental drafts 218
12.4.3. Interventions of the quality specialist in the creation of drafts of the innovation 221
12.4.4. The interventions of the quality specialist for the creation of drafts of future products or services 226
12.5. The missions of the quality specialist in impact studies on competitive quality fuel 232
12.5.1. The role of the quality specialist in the characterization of future products or services 235
12.5.2. The role of the quality specialist in identifying future clients 237
12.5.3. The role of the quality specialist in the characterization of future uses of the innovation 240
12.5.4. The role of the quality specialist in carrying out prospective studies 243
12.5.5. The role of the quality specialist in carrying out tests on the drafts of future products or services 247
12.6. The missions of the quality specialist in the feasibility studies 253
12.6.1. The role of the quality specialist in conducting the initial feasibility tests of the future products or services 256

12.6.2. The role of the quality specialist in identifying future stakeholders in the project. 258
12.6.3. The role of the quality specialist in the feasibility studies on exchanges with the future stakeholders in the project. 259
12.6.4. The role of the quality specialist in carrying out performance tests of future products or services incorporating the innovation 263
12.6.5. The role of the quality specialist identifying stakeholders in the future BU 265
12.6.6. The role of the quality specialist in studies on the ability of the company to produce future products or services 267
12.7. The role of the quality specialist in equipment design 270
12.7.1. The role of the quality specialist in the construction of the innovation processor 273
12.7.2. The role of the quality specialist in the design of prototypes of the production system for future products or services 275
12.7.3. The role of the quality specialist in the intellectual protection of the innovation 277
12.8. The role of the quality specialist in assisting decision-making. 278
12.9. Conclusion. 285

CHAPTER 13. QUALITY CULTURE IN PROJECT TEAMS. 287

13.1. Introduction. 287
13.2. The inherent visions of quality 288
13.3. The aptitudes for producing quality 290
13.3.1. Individual aptitudes. 290
13.3.2. Collective aptitudes 294
13.4. Methods for acquiring aptitudes for producing quality. 297
13.5. Conclusion. 298

APPENDICES 299

APPENDIX 1 301

APPENDIX 2 309

APPENDIX 3 319

BIBLIOGRAPHY 335

INDEX 339

Preface

This book is aimed at two categories of readers.

The first category of readers is made up of directors, innovation project managers in companies with permanent Research and Development departments, project team members, project promoters undertaking an innovative approach which will form the basis of a new company, public stakeholders responsible for promoting innovation, anyone wishing to understand the basics of using quality to achieve success in innovative approaches and so on.

These readers should consult Chapters 1, 2 and 13, and the first and second sections of Chapters 3–11. They can easily skip Chapter 12, which is specifically aimed at the second group of readers.

The objective, as far as these readers are concerned, is to help them to understand the power of a quality approach which is integrated into innovation processes so that their development is able to draw from all possible energy sources, thus ensuring their success.

The second category of readers consists of directors of quality departments, members of quality departments who assist project managers, Master's degree students specializing in quality management, consultants specializing in quality who work in cooperation with project promoters and so on.

These readers can read this book in its entirety and, for the more technical aspects, refer to the works cited in the bibliography.

The objective here is to help them to grasp the concepts which are at the heart of every quality approach integrated into an innovative process, and to offer guidelines for actions to pursue when carrying out such approaches, showing the link between the fundamental aspects of each quality production approach in a process of creating added value.

I would like to thank all the companies with which I cooperated in their innovative approaches, allowing me to draw from their experiences in order to write this book.

The areas where innovations are developed are obviously highly protected. A high level of confidentiality must be respected. It is for this reason that no concrete or very precise examples are given to illustrate the methodological developments which are discussed. This is also why the companies which rely on quality to optimize their chances of success in innovative approaches do not wish to be identified.

Many of these companies have hesitated for some time before introducing a true quality approach within their departments responsible for promoting the creation and development of innovations. Their experiences in relation to quality approaches integrated into production processes suggest that the restrictive procedures inherent to such approaches tend to be rejected by those stakeholders concerned and encroach on the freedom necessary to foster innovation.

Companies felt that innovating means taking risks, while a quality approach is designed to avoid such risks; innovating means focusing on an uncertain future, looking out for all sources of information which could provide added value, while a quality approach encourages us to look back in order to guide decision-making; innovating means taking an action that could not be reproduced, while a quality approach is designed to identically reproduce a predefined process.

All these views, while perfectly understandable, are clearly opposed to a quality approach in the innovation process.

These companies were aware of the fact that innovating in an ever more complex environment was becoming risky, that a continually growing number of demands as well as the behavior of socioeconomic stakeholders needed to be taken into account, and that the action of only one of these stakeholders could lead to the failure of an innovation.

"Incubators" were the first to ask us for advice on how to use quality as a means to optimize exchanges with these various stakeholders, without weakening the creative thrust of the teams responsible for the development of innovations. Companies operating in the health sector quickly followed.

Then, the large companies with "research and development" departments also started to move in this direction. Directors of these departments did not want the field of quality services to encroach onto their own territory. A complete reconfiguration of the field of action of "quality specialists" was needed for them to be accepted within these structures.

Today, these companies could not do without their "creative quality specialists", rather than the "regulatory affairs managers", in order to optimize the chances of success for their innovative approaches.

Pierre MAILLARD
December 2014

Introduction

Global economic growth is plateauing because overall supply is growing faster than demand. Global growth is being led by those countries whose economic development is recent and whose driving force is a supply which impacts exports by manipulating prices. This leads to a slower creation of internal wealth which is concentrated in a small percentage of the population. Therefore, the overall internal demand evolves more slowly than the supply, which is focused on high-end services usually exported toward the so-called "industrialized" countries.

These industrialized countries have been invaded by a lost-cost supply responding to a demand from a large population consuming a continually growing quantity of products and services.

All industrialized countries are making a special effort to revive their economic activity through innovation. They are, therefore, attempting to create new value which differentiates them from the production driven by the strengths generally present in emerging countries. They use innovation to find answers to the problems in society, and to benefit as quickly as possible from the scientific research which represents a key reserve of competitive growth.

Innovation is therefore at the heart of our civilization's economic and social concerns. All our hopes for finding solutions to our problems or for satisfying our ambitions lie in innovation.

The main strengths of emerging markets are the following: a significant potential market for traditional products and services, low labor costs, a highly attractive location for investors and strong development potential for companies in industrialized countries. These companies are happy to quickly transfer their expertise for power: gaining access to these markets, producing locally to maximize profit margins and rapidly reaching a global size which is comparable to that of their main competitors.

The economic assets of industrialized countries are their culture, their ability to capitalize on extensive expertise, a significant financial platform for investing in innovation, and higher purchasing power that opens markets not only to traditional products and services which can be made in emerging countries but also to innovative products which they produce themselves. These markets can serve as tests to evaluate the chances of economic success for a given innovation. They can promote the emergence of an attractive image for innovative products which will then be used, through fashion trends, to access markets in emerging countries more quickly by initially positioning them in high-end sectors and gradually allowing them access to sectors with greater numbers of customers, thus justifying the construction of local production facilities.

The fundamental problems in our society are centered on the protection of the environment, energy saving and finding better uses of renewable energy, recycling raw materials, managing migratory flows caused by socioeconomic inequality and the performance of information and transport methods, the growth of the human population and the aging of this population.

For example, without therapeutic innovation, the lifespan of human beings would not have been lengthened to such an extent. Without technological innovation, modes of transport would not be so reliable, quick and comfortable. Without technological innovation, the number and diversity of products would not have reached their current level, and so on.

> *Innovation accelerates change and innovation is needed to control or make use of this acceleration.*

Nevertheless, paradoxically, innovation can cause instability, and the consumer becomes used to it or even expects it. They are increasingly less attracted by innovation, and this innovation must be incorporated into systems that are increasingly more complex.

Consequently, the risk of failure becomes increasingly great.

This new economic dynamic highlights the importance of introducing quality approaches into innovation processes to increase the chances of success for companies wanting to develop through innovation.

However, researchers and creators fear that these approaches may further reinforce the restrictions which limit their freedom.

The tests which we carried out in various companies show that if we are not trying to reproduce quality approaches introduced into mass production processes, and if we design specific approaches that guide and stimulate research or creativity, then we can then considerably increase the chances of these innovations succeeding.

Competitiveness clusters, which in France are catalysts for innovation, should be the main promoters of this new use of quality to increase the chances of success for projects they support.

Unfortunately, it can be noted that currently the tools of quality approaches which ensure the success of innovations are not central to the concerns and focus of these competitiveness clusters. The directors and stakeholders of these clusters simply see the "normative technocracy" of quality as too often restricted by the main sources of information which are today made up of regulatory and certification bodies.

There are other ways of using quality to ensure companies' development and the success of their innovations, and we deemed it necessary to support the dissemination of these other quality practices.

This is the goal of this book. It presents the main aspects which characterize these new quality practices in innovative approaches.

The contributions of quality mainly focus on the following objectives:

– to best ensure that an innovation:

- will create and strengthen the desired links between the company and the sectors of the market targeted;

- will create the purchasing value expected of products and services which integrate the innovation by stimulating certain perceptions that clients have of these products and services;

- will provide the resources that internal and external investors and producers of products or services will require in order to produce them;

- will comply with the regulatory and security demands of clients and institutions of countries in which these market sectors are located.

We will look in more detail below at the concept of "purchasing value". This can be expressed as a percentage of purchasing power of a market sector which is dedicated to purchasing products and services which promote innovation.

The role of quality is therefore fundamental in the implementation of company policies aimed at using innovation to increase competitiveness.

This role is necessary to give confidence to those financing innovation and decision makers.

The management of quality integrated into the delivery of innovative projects focuses on two competitiveness factors: controlling risks of failure caused by breaking with previous production and consumption practices, and optimizing the access to markets created by innovation.

Every project designed to introduce an innovation to a product or service brings with it a number of uncertainties which could entail various risks of failure. In order to accept a commitment of significant resources in this type of project, it is necessary to include in their management an effective regulation system which is able to identify these risks as soon as possible and avoid them having an impact on the economy of the project, while taking full advantage of the creative potential of the project team to attract the markets. This is the natural role of a quality system integrated into an innovation process. It is a case of finding and maintaining this "dynamic equilibrium" between the best possible use of value creation opportunities and controlling risks of failure.

In this book, we will greatly elaborate on the specific concept of "dynamic equilibrium" integrated into innovation processes.

In order for this very specific type of quality system to be effective, it must incorporate a number of specific characteristics which are detailed later.

It can be noted that the efforts made by most industrialized countries to promote innovation mainly consist of creating public funds designed to provide financial assistance to all innovative project promoters; creating structures designed to pool the skills necessary for fostering innovation, especially on the level of small and medium enterprises (competitiveness clusters, incubators, etc.); and encouraging research centers to target the development of their results toward technological innovation (development of transfer centers).

The public policies are mainly focused on "technological" innovation.

But technological innovation.

Technological innovation is not an end in itself in order to build the economy of a developing country. It is simply an essential element in part of the wider process of developing technological innovation which is able to respond to a significant and resolvable need of a market which it has helped to create.

It is therefore necessary for public policies to work alongside innovations in services, and the implementation of specific quality approaches in innovative projects, which are essential to increase the innovations' chances of success both in economic and social terms.

Nowadays, we often hear that it is easier to innovate in products than in services, while the non-physical nature of services should instead make them more open to innovation. This observation is mainly due to the essential role of man in the production of services, and the difficulties in breaking with individual and collective behavior and attitudes.

The non-physical nature of these components in the production of innovations is also one of the main causes for the low level of interest shown by policies designed to create growth through innovation.

It is difficult to communicate to economic authorities the importance of services and quality approaches integrated into innovation production systems in order to ensure the creation of value leading to greater and more sustainable profits.

Nevertheless, all recent studies show that the importance and the duration of an innovation's social and economic impact mainly depends on these two components of its production.

The complexity of the situations in which innovations are developed today, the current coverage of most of the markets' fundamental requirements with existing products, the rise of consumerism leading to increasingly more demanding clients, and the role of the media in creating fashions, with images of innovative products to create new needs and which are the basis of returns on investment, are all factors which justify the key role of services and quality approaches in the success of technological innovations.

It is already the case, but will be even more so in the future, that the product at the vanguard of innovation will only be used to sell services, and the recognition by the market of the quality of the innovation will be the main factor in ensuring returns on investment.

Policies, mainly focused on technological innovation, risk inciting significant investment which, in the medium term, will not produce the hoped-for effects for the revival of developed countries' economies.

In order to change these attitudes, it is necessary to provide policy-makers and innovation producers with the means to better understand the way in which quality approaches integrated into innovative projects and services associated with innovative products enhance such innovations and increase their chances of success. It is also necessary to help them to take on and use these key components in producing value in the current social and economic context.

This is the aim of the conceptual and methodological accounts given in this book.

1

The Innovation Concept

1.1. The characteristics of an innovation process

An innovation process is a production activity which takes place prior to the design process.

We would define such a process in the following way:

> *An innovation process is the gradual materialization of new ideas guided by the projection of a very uncertain future, in terms of its methods of use and the returns on investment that they will produce.*

The process starts with a meeting of "ideas" and "needs" of a population that wishes to carry out a given activity.

The "ideas" refer to using scientific, technological or methodological results from fundamental or applied research, or certain components of products or existing services, in order to create new "features" which respond to specific areas of society not currently covered or not yet catered for well enough. The targeted population of beneficiaries must be sufficiently large and have sufficient financial resources to ensure the expected returns on investment.

These new "features" often lead to changes in the organization of the population which receives the innovation. They can change habits and destabilize the current order. Moreover, satisfying needs by "supplying a service" which includes an innovation can create further needs. A consumer process then develops, and initiates the development of new economic exchanges.

EXAMPLE 1.1.–

Remote cardiology is the result of an idea based on using new information technology to create a permanent connection over large distances, with the use of satellites, between a patient suffering from a serious cardio-vascular illness and a medical team responsible for the patient's treatment.

The need is clear. The information technology exists. The idea was to connect this technology with medical treatment.

The creation of an innovation process brings together the human resources within a company, which promote the idea, understand the needs and have the skills necessary in the field of information technology.

The way in which the innovation is constructed is dependent on the material resources used (cardiac electrodes, defibrillator, software contained within the defibrillator, transmitter, external switch controlled by the patient, methodology of purchasing satellite services, ergonomic information receiver for cardiologists) and on the way the features of the service are defined (providing the cardiologist with accurate and reliable information for to give a remote diagnosis, locate the patient, get in touch with him quickly, etc.).

The service includes a connection between the patient and a permanent remote supervision service for making a pre-diagnosis. He transmitted the information at a team of caregivers on the spot for make-out the diagnostic and care.

This new supply of services created through innovation in turn gives rise to other needs regarding remote treatment for the evolution of other illness which require urgent intervention.

As mentioned above, the “promoters” of the ideas and “needs” are part of the human resources responsible for the development of the innovation.

“Innovation” and “innovation process” are terms which are frequently used to express the evolution of the resources that human develops in order to satisfy his needs.

Our goal is to offer a methodology designed to integrate a quality process into an innovation process in order to ensure, over time, the most cost-effective competitiveness possible: from the project which creates the innovation and makes it economically viable, to the production system for the products and services which will benefit from this innovation and, finally, to the products and services when they are brought into the market.

These particular quality processes are known as “competitive quality processes”.

An innovation process can be represented in the following way:

We voluntarily end the innovation process before the decision is taken, by the company management, to implement the innovation by undertaking product or service designs which will incorporate the innovation, and the design of the production system. Here, we depart from the innovation process and start the design process. It is still considered in some companies that the innovation process only ends with an "innovation on the shelf".

> An innovation process consists of the following steps:
>
> – deciding to foster new ideas to help the progress of the company;
>
> – bringing about the emergence of an idea which may provide a response to a need of society;
>
> – briefly characterizing products or services which could benefit from the innovation;
>
> – identifying possible clients for these future products or services;
>
> – characterizing future uses for products or services which would incorporate the innovation;
>
> – creating experimental drafts which represent the idea as accurately as possible;
>
> – conducting initial feasibility tests on the future products or services;
>
> – deciding to initiate prospective studies on how to use the idea to benefit the company;
>
> – conducting prospective studies on possible socioeconomic uses for the innovation, and its impacts on the company and its position;
>
> – deciding to guarantee the ability of the company to mobilize the resources necessary, internally and externally, in order to fully realize its innovation process;
>
> – identifying future stakeholders in the innovation project;
>
> – conducting studies on the nature and feasibility of exchanges with future stakeholders in the project necessary for it to be fully realized;
>
> – deciding contractual issues with these stakeholders, and building a processor to carry out the innovation process;
>
> – building a processor responsible for carrying out the innovation process and an external communication network;
>
> – deciding to progressively design drafts which give a more accurate representation of the idea;
>
> – progressively designing these drafts to accurately represent the configuration of the innovation;
>
> – deciding to carry out prospective studies on the socioeconomic impacts of the future products and services which will incorporate the innovation;

– progressively designing drafts which give a representation of future products and services to incorporate the innovation, by describing, in particular, the future "interfaces" of the innovation with other components of these products and services;

– conducting tests on the drafts of future products or services, at various steps in their production, on future clients and on all future stakeholders in the production system for the products and services;

– conducting performance tests on future products or services;

– creating a rationale to aid decision-making, which examines the probable socioeconomic impacts of implementing the innovation;

– deciding to guarantee the ability of the company to adopt a production system for the future products and services;

– more precisely identifying stakeholders who will contribute to the development and production of future products and services which will incorporate the innovation;

– designing drafts of the production and distribution systems for the products and services;

– conducting studies on the ability of the company to design production and distribution systems for the products and services which will incorporate the innovation by taking into account the configurations of the various definitive drafts of the innovation;

– creating a rationale to aid decision-making, which details the feasibility of implementing the innovation;

– deciding to validate the innovation;

– examining the steps to take to ensure the intellectual protection of the innovation.

Box 1.1. *Characterization of an innovation process*

COMMENT 1.1.– Each of the previous steps can be carried out several times. An innovation process is carried out "in loops". It is possible, at any time, to question ideas, observations or previous decisions.

We detail below the meaning of the terms used.

With this representation of an innovation process, it is understood that

An innovation is an original way of developing a socioeconomic activity by attempting to find a new way of satisfying the needs of the society.

There are various definitions for the concept of innovation. It would certainly be interesting to recall them by trying to identify and characterize the points of view which form the basis of such definitions. We leave this work to one side in order to focus on the crux of our question regarding the characterization of a competitive quality process integrated into an innovation process.

An innovation process can take on very different forms depending on the context in which it is carried out (size of the company, sector of the economy, creation of a company through innovation, etc.).

A quality management specialist does not have the competency to directly influence project leaders in choosing the most optimal methods to carry out innovation processes. They must only focus on those actions which make the best possible use of the specific production mechanisms for "quality" within the innovation process and prevent them from hindering the creativity which is at the heart of its added value. The specialist's role is to design and implement a quality process which is integrated into the innovation process, providing it with the most effective support to help it achieve economic success. However, the information produced by this quality process can help improve the performance of the innovation process as it proceeds or when it is reproduced, when the configuration of the production system is redefined.

To be able to integrate a competitive quality process into an innovation process, we need models which represent the various concepts which are involved.

In the previous model describing an innovation process we can observe five main categories:

– creative research;

– decision-making;

– impact studies;

– feasibility studies;

– organizing production processes.

Box 1.2. *Main categories which make up an innovation process*

Creative research is a highly fragile area of the innovation process. It is greatly influenced by the conditions in which it is carried out and must be afforded a considerable amount of freedom. It is creative research which produces the most significant portion of added value within the innovation process.

Creative research are involved in the following stages of the innovation process:

– bringing about the emergence of an idea which may provide a response to a need of society;

– creating experimental drafts which provide a more accurate representation of the idea;

– progressively designing drafts to accurately represent the configuration of the innovation;

– progressively designing drafts which give a representation of future products and services to incorporate the innovation, by describing, in particular, the future "interfaces" of the innovation with other components of these products and services.

Box 1.3. *Creative research*

Impact studies can be described in the following way:

Impact studies are involved in the following stages of the innovation process:

– briefly characterizing products or services which could benefit from the innovation;

– identifying possible clients for these future products or services;

– characterizing future uses for products or services which would incorporate the innovation;

– conducting prospective studies on possible socioeconomic uses for the innovation, and its impacts on the company and its position;

– identifying stakeholders who will contribute to the development and production of the future products and services which will incorporate the innovation;

– conducting tests on the drafts of future products or services, at various steps in their production, on future clients and on all future stakeholders in the production system for the products and services.

Box 1.4. *Impact studies in the innovation process*

These studies attempt to predict the socioeconomic consequences of the innovation. These impacts are predicted by studying similar cases or by carrying out tests based on simulations or experiments. In this way, the predictions of impacts can be made credible.

As for feasibility studies, these can be characterized in the following way:

Feasibility studies are involved in the following stages:

– conducting initial feasibility tests on the future products or services;

– identifying future stakeholders in the innovation project;

– conducting studies on the nature and feasibility of exchanges with future stakeholders in the project necessary for it to be fully realized;

– conducting performance tests on future products or services;

– conducting studies on the ability of the company to design production and distribution systems for the products and services which will incorporate the innovation by taking into account the configurations of the various definitive drafts of the innovation.

Box 1.5. *Feasibility studies in the innovation process*

These studies attempt to anticipate how to capitalize on opportunities, and how to account for restrictions, in order to achieve the socioeconomic objectives which justify the investments to be made.

It is essential to anticipate the ways that the innovation could be used in the future in order to prevent directors from leaving the innovation "on the shelf".

Some stages of the innovation process are in the field of organization, and they are the following:

> The *organizational stages* are the following:
>
> – building a processor responsible for carrying out the innovation process and an external communication network;
>
> – creating a rationale to aid decision-making, which examines the probable socioeconomic impacts of implementing the innovation;
>
> – designing prototypes of the production and distribution systems for the products and services;
>
> – creating a rationale to aid decision-making, which details the feasibility of implementing the innovation;
>
> – examining the steps to take to ensure the intellectual protection of the innovation.

Box 1.6. *Organizational stages*

Finally, decision-making stages are key at every step of the process. They can be identified in the following way:

> The *decision-making stages* are:
>
> – deciding to foster new ideas to help the progress of the company;
>
> – deciding to initiate prospective studies on how to use the idea to benefit the company;
>
> – deciding to guarantee the ability of the company to mobilize the resources necessary, internally and externally, in order to fully realize its innovation process;

– deciding contractual issues with these stakeholders, and building a processor to carry out the innovation process;

– deciding to progressively design drafts which give a more accurate representation of the idea;

– deciding to carry out prospective studies on the socioeconomic impacts of the future products and services which will incorporate the innovation;

– deciding to guarantee the ability of the company to adopt a production system for the future products and services;

– deciding to validate the innovation.

Box 1.7. *Decision-making stages in the innovation process*

These stages are always sensitive because the risks associated with them often involve high criticality levels which are not always accurately estimated.

It is necessary to segregate these various operations because they rely on different logics and means. It is generally impossible to offer a common planning model for these various operations in spite of there being some clear precedence relations. Each situation requires a specific plan.

An innovation process must remain extremely flexible to quickly adapt to uncertainties and new opportunities which emerge during the process. This is characterized by:

– a rough initial characterization of the final product;

– unrestricted evolution of this characterization as the innovation production process progresses;

– uncertainty regarding the configurations of successive drafts representing the evolution of the innovation process;

– the necessary evolution of the "dynamic architecture" of the system which follows the production process according to how results are used to benefit the innovation;

– imprecision and evolution of guidelines for the management of the process;

– the necessary pooling of information resources and decision-making partnership;

– uncertainty regarding the impact on future users' satisfaction;

– uncertainty regarding investments to be made.

These specificities may seem to imply that it is dangerous for a company to commit to such projects. In reality, in the current economic context, a company established in an industrialized country which does not innovate has little chance of survival, except if it is positioned in a strategic niche not threatened by competition for emerging countries, or if it has such a strong historical reputation that it has become timeless. In this particular case, the company is well protected against the competition, and must instead evolve slowly and gradually so as not to destroy those products and services which have earnt it such a reputation.

Such vagueness, uncertainty, arbitrariness and imprecision mean that the innovation process must be very flexible and reactive in order to constantly adapt to the often unpredictable information which is produced to act as a guide in the process.

The specificities of the innovation process show that significant risks are involved. Nevertheless, it is generally the only way to put products on the market which are more attractive than those of competitors, and to encourage consumers to spend a larger portion of their purchasing power to obtain them.

These specificities show that it is impossible to apply the traditional mechanisms for controlling risk within a process in order to bring it criticality down to acceptable levels. These mechanisms rely on methods which attempt to reduce the uncertainty at the source by identically reproducing what gave the best results in the past. They abolish levels of freedom necessary for individual and collective creation which are the basis of all innovations.

Finally, it is impossible to give a definitive *a priori* definition of the "dynamic architecture" of the innovation production systems. The emergence of new ideas throughout the innovation process shapes its evolution and can considerably modify its initial "dynamic architecture". Those responsible for carrying out the innovation process must always be able to modify this initial definition in order to successfully realize the production of the innovation, according to the opportunities which emerge from the creation of new ideas during the previous stages of the innovation process, and the new risks arising when these opportunities are taken. Such management of the innovation process is particularly sensitive because it requires a number of variables which are often not initially included in the supervision plans to be observed, while at all times preventing any restriction of the actions promoting the creativity which produces the innovation's added value.

This management requires a high level of vigilance, an ability to rapidly understand and interpret the signs which indicate the emergence of unexpected risks and a "culture" within the company which allows for the rapid estimation of "the hope of profits created from opportunity of implementing an idea" and "the criticality of predictable risks before it is implemented, or indicated during its implementation", in order to undertake corrective actions as soon as possible that lower the criticalities to acceptable levels while minimizing the loss of hoped-for profits.

The management architecture for the production of an innovation must be in "matrix" form, as with any complex project. The "horizontal" organization of a management system, which ensures the direction of the project, must consist of a considerable ability on the part of the company's general management to delegate responsibility. This delegation must be compensated by precise management of the layout of intermediary results presented during project review stages. Regulation of the innovation process cannot be self-governed, as with other production processes. Members of the project team producing the added value in the innovation process must focus on the creation of this value. The project manager must most often delegate responsibility to the quality service. However, the partial results of this regulation must be scrupulously addressed during project reviews, and alerts must be followed by responses as soon as possible.

Broadly speaking, the models that we propose are based on the fundamentals of "general systemics".

1.2. Review of basic conceptual definitions in "general systemics"

Above, we have described a number of concepts which are consistently at the heart of developments which will follow, and it is useful to be able to define them accurately.

1.2.1. *The concept of a process*

The term "process" is given to any series of events linked over time which has the following properties:

– all events contribute to the same end result;

– all events start at a moment called the "initial time" of the process;

– all events take place before a given time called the "end point" of a process;

– the series of events linked over time obeys some kind of organizational framework;

– every process can be broken down into component sub-processes.

1.2.2. *The concept of a processor*

The term "processor" is given to a dynamic system which includes human resources. It is activated by these human resources and its activity is led by a specific "managerial" organization.

COMMENT 1.2.– Every process is carried out by activating a processor. A processor can carry out a number of different processes.

1.2.3. *The concept of functionality*

A "functionality" is the ability of a processor to transform a system when it is activated. The system can be the processor itself.

1.2.4. *The concept of a function*

A function is the transformation of a system.

COMMENT 1.3.– By activating a functionality of a processor, a function is carried out.

1.2.5. *The concept of configuration*

The "configuration" of a system is the architecture of its resources which may be human, material, information-based, organizational, financial and temporal, coupled with its "functional" potential (a collection of its functionalities).

1.2.6. *The managerial architecture of a processor*

The human resources of a processor have a specific organization which influences other resources (material, information-based, financial and temporal).

In general, three main components of this architecture can be distinguished:

– "pyramid" managerial organization;

– "horizontal" managerial organization;

– support services.

"Pyramid" managerial organization, at the company level, is responsible for its overall "dynamic equilibrium". This is what determines the socioeconomic position that the company must attempt to occupy. This is what distributes the resources among the various processors. This is what sets the objectives that each processor must reach. This is what fulfills the company's legal and administrative responsibilities. This is what controls the way in which internal relations between the processors and communication between the processes are carried out. It is partially present in every "processor" and is visible in the "hierarchical" levels within the processor. The organizational figure of the company's management usually shows this "pyramid" structure. A company can have several hierarchical levels. The number of levels depends largely on the size of the company.

"Horizontal" managerial organization is specific to each process. It is also known as the "operational management" within a company. This is the organization directing the creation of added value for the company which will be visible to its clients and will therefore allow the company to develop exchange relationships with its external stakeholders in order to acquire and build up its reputation. It is directed by the process leadership, and brings together the human resources supplied by "professions" or "support services" to carry out the actions in the process. The features which characterize the authority, responsibility and assistance of this leadership to its human resources are determined by the contracts that it agrees with the representatives of the "professions" or "support services". In general, three levels can be distinguished in this type of organization:

– process leadership;

– the representative of each working group carrying out a task in the process;

– each individual in the working groups.

The support services are "cross-cutting professions", which are involved in all activities within the company. They help all stakeholders in a company carry out specific cross-cutting activities such as supplying resources, coherence and efficiency of the interfaces between the various company departments, the coherence of its activities and the overall conformity with specifications imposed by regulations or laws.

Among others, they bring together the "personal" function, the administrative function, the "quality" function, the function responsible for the maintenance of material resources, the "marketing" function and so on.

1.3. Evaluation criteria for the success of an innovation

It would be tempting to say that the success of an innovation can be measured by its effect on fulfilling the company's policy. However this evaluation criteria demand too much time to ascertain whether an innovation has been successful and whether an innovation can, because of its success, encourage the company to change its policy.

The main criteria used to evaluate the success of an innovation process are as follows:

The main evaluation criteria for an innovation process include:

– the evolution in market uptake in terms of volume;

– the speed and size of returns on investment;

– the evolution in the purchasing value given by the market to the products or services which incorporate the innovation;

– the evolution in the attraction to the innovation compared to that of competitors' products or services;

– the nature and extent to which the innovation provided an answer to society's evolving needs;

– the level of contribution to the enhancement of the fundamental image of the products or services which incorporate the innovation, or to the fundamental image of the company;

– the level of enhancement of the "expertise" and culture of the company;

– the ease of finding and using this new "expertise" necessary for the production of new products and services;

– the time taken to introduce the products or services which incorporate the innovation into the market;

– the contribution of the innovation to the fulfillment of the company policy.

Box 1.8. *Criteria for a successful innovation*

There are certainly additional criteria, and some companies find it necessary to personalize these criteria.

These criteria can be subject to a predictive estimation process at the start of the innovation process. This estimation is periodically re-evaluated during the process to guide decisions or, after decisions are taken, with regard to continuing or terminating the process or correcting the direction of the process. At the end of the process, a final estimation can serve as an objective to evaluate the success of an innovation during the lifespan of a product or service which incorporates it.

In general, it is possible to allocate one or several quantifiable indicators to each of these factors.

The predictive nature of the innovation's performance evaluation is traditionally represented by attaching to each quantifiable indicator an estimation of its value in terms of "confidence intervals".

This estimation carries risks for the stakeholders investing in the innovation project. These risks are linked to the uncertainty of the information used in its formulation, and to its predictive nature. Each stakeholder must assess their own "risk benefit ratio" according to the evolution of the estimations as the project progresses.

1.4. Drivers of socioeconomic exchange for an innovation process

An innovation process is a project which relies on a great number of exchanges between the central system which creates the process's added value and its environment. The central system is a processor, which has human resources often called the "project team". In small businesses (SMEs), this team is sometimes very small with just one or two people. As soon as there is more than one, a project leader is nominated. He has full operational managerial delegation in his relations with the other members of the project team. The project team generally changes very little during the innovation process. All other components of the processor can evolve according to the needs which emerge as and when the configuration of the innovation changes.

To introduce the model for drivers of socioeconomic exchange, it is necessary to define the meaning of the concept "stakeholder".

The "stakeholder" of a processor is any socioeconomic agent which carries out exchanges with the processor.

Box 1.9. *The concept of a stakeholder*

The project team is a particular class of stakeholders in the process. It needs to enter into relations with a large number of other stakeholders to supply the processor with resources.

Apart from the architecture of the processor's human resources which changes little during the innovation process, the other components of its configuration can significantly evolve as the project advances. These changes occur through exchanges between the processor and its environment.

In a general way, the organization of the various stakeholders in a processor can be modeled by the concept of "strategic areas of activity".

A strategic area of activity consists of:

– a service package which the company offers to its clients;

– a company processor which produces this service package; (Business Unit – BU)

– a family of stakeholders who make up the clientele for the service package; (Market)

– a family of stakeholders who produce certain components in the service package within the "BU"; (Workers)

– a family of stakeholders external to the "BU" who provide certain components for the service package, or resources which are then processed by the company in order to produce these services; (Providers)

– a family of stakeholders who invest in the company hoping to make a profit from the production of these services; (Investors)

– a collection of surrounding parties who are connected to any of the above stakeholder

Box 1.10. *The concept of strategic areas of activity*

Hence, the following figure results:

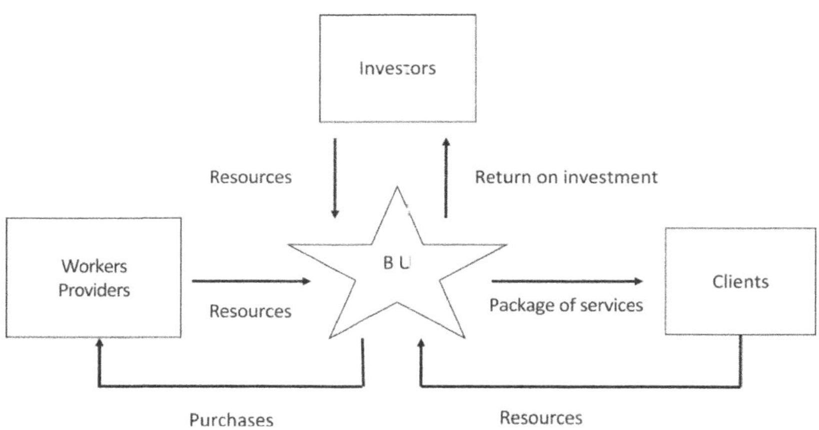

Figure 1.1. *Strategic area of activity*

The driver of socioeconomic exchanges in the innovation process can be described in the following way:

> The driver of socioeconomic exchange of the innovation is made up of:
>
> – a processor built around a project team and which produces the added value;
>
> – an area of strategic activity connected to this processor;
>
> – a future area of strategic activity connected to the "BU" which can economically benefit from the innovation and which, in particular, contains the class of clients in the previous area of strategic activity;
>
> – a network and a flow of exchanges between the processor and its stakeholders;
>
> – a communication network and flows of information between the processor and its surrounding parties;
>
> – networks of exchanges of information, products, services and money between its stakeholders, or between its stakeholders and certain connected parties.

Box 1.11. *The concept of a driver of socioeconomic exchange*

In order to undertake a competitive quality process, it is first necessary to describe these significant components which make up the driver of socioeconomic exchanges of the innovation.

It can be noted that the driver of socioeconomic exchanges consists of two strategic areas of activity: one which connects all stakeholders to some aspect of the innovation process, and another which will connect all the stakeholders to be involved in the future exchanges which the company must undertake to benefit from the innovation. In the second case, the hypothetical future projection affects the way the innovation process is carried out.

The second strategic area of activity is a model of the organization of exchanges between the various levels of clients in the first strategic area of activity.

The first strategic area of activity can be broken down in the following way:

> The strategic area of activity connected to the innovation processor is made up of:
>
> – a package of services that the processor offers to members of the company's BU which will benefit economically from the innovation; (Project offer)
>
> – the innovation processor itself; (Production system for the innovation process)
>
> – the strategic area of activity of the BU which arranges the clients of the innovation processor;
>
> – the family of stakeholders who produce the package of services; (Project team)
>
> – the family of stakeholders, external to the project team, who provide certain components of the service package, or resources which are then processed by the project team to produce these services; (Project providers)
>
> – the family of stakeholders who invest in the project hoping to benefit economically from the innovation; (Project investors)
>
> – a collection of parties connected to any of the above stakeholders and which have an indirect effect on the project.

Box 1.12. *The strategic area of activity of an innovation process*

The second strategic area of activity can be described in the following way:

> The strategic area of activity of the future BU which will utilize the innovation is made up of:
>
> – the service package which the BU will offer to its clients by utilizing the innovation; (Supply)
>
> – the processor which will produce the package of services; (BU)
>
> – the family of stakeholders who will make up the future clients of the service package; (Potential market for the innovation)
>
> – the family of stakeholders who will produce certain components of the service package in the "BU"; (Workers)
>
> – the family of stakeholders external to the "BU" who will provide certain components for the service package, or resources which will then be processed by the company in order to produce these services; (Providers)
>
> – the family of stakeholders who will invest in the BU hoping to benefit from the production of the services; (Investors)
>
> – a collection of related stakeholders who are connected to any of the above stakeholders.

Box 1.13. *The strategic area of activity of a business unit utilizing an innovation*

COMMENT 1.4.– It may seem that the strategic area of activity of the innovation processor has a particularly complicated organization with its clients. In reality, this complexity is present for the majority of service production processors. They generally have a stream of potential clients.

1.5. Clarifications on certain actions in an innovation process

1.5.1. *Managing the configuration of an innovation process*

One of the fundamental rules of managing an innovation process is commonly referred to as the "*management of its configuration*".

It is characterized by the periodical organization of project reviews which consist of answering and recording the responses to two questions:

– What information has been produced since the last project review which contributes to providing:

- an update on the latest configuration of the draft, which embodies the evolving ideas at the source of the innovation, intended to guide the future design of the product or service to incorporate the innovation;

- an update on the latest configuration of the processor draft which will create the future product or service;

- an update on the probable impacts on relations between the company and given stakeholders;

- the risks of not achieving these impacts and an estimation of their criticality levels.

– What actions should be earmarked in order to supplement the configuration of previous drafts, to refine the impacts on relations between the company and its stakeholders, and to reduce the risks of not achieving such impacts to acceptable levels?

Managing the configuration at each stage of the project includes: the progress of the evolution process, the choices which steer the continuation of the process and the reasons for these choices, the objectives to fulfil, the resources to assemble and the scheduling of tasks to carry out.

Those involved in an innovation process often struggle to carry out this configuration management properly. They feel that they are wasting time, even though they are convinced of the merits of such actions. Progress can only be tracked by using a knowledge base acquired previously, and by tracing the mechanisms which allowed this knowledge to be accumulated.

1.5.2. *Creative research*

An innovation process becomes official in a company when an order is given from the management of that company. This order is given through a proposal which consists of:

– an idea which could satisfy a need within society;

– a prospective vision of how to design a product or service incorporating the idea which could be used by the population;

– an approximate prediction of the benefits that the company could receive as a result of the socioeconomic impacts of the innovation.

This proposal provides the input data for the creative research process.

As discussed above, creative research consists of the following stages in the innovation process:

– bringing about the emergence of an idea which may provide a response to a need of society;

– creating experimental drafts which provide a more accurate representation of the idea;

– progressively designing drafts to accurately represent the configuration of the innovation;

– progressively designing drafts which give a representation of future products and services to incorporate the innovation, by describing, in particular, the future "interfaces" of the innovation with other components of these products and services.

The stakeholders involved in this task have evolving resources. They are the only ones responsible for presenting the updates to the project backer. These updates are given during the project reviews which validate the progress at each stage in the creative research process. It is possible for each stage not to comply with previous predictions since these are simply working hypotheses. Innovation processes are not set in stone. Managers have a significant amount of freedom to modify their direction at any time according to opportunities or problems which arise during the process.

A program can consist of several areas of "creative research" which open up a relatively wide field of research, while nevertheless remaining fixed on the same objective.

This objective generally remains vague. It is an "idea" which becomes more precise as the process advances.

In summary, it is possible to characterize the architecture of a creative research processor in the following way:

– the end result is an "objective" expressed by an approximate idea;

– the division of sub-processes is structured into different areas of "creative research";

– each sub-process becomes more precise as the process advances;

– the processor has a variable geometric layout arranged around a "hub" of interactive resources which remains unchanged and constitutes the basis of the innovation driver;

– each "hub" has a stable "horizontal" managerial organization led by a project manager;

– some of the hub's resources and the flexible components of the processor are provided by specialized departments within and outside the company;

– these departments are reservoirs of resources which can supply the processor, or constitute sub-processors, depending on the project manager's decisions;

– when these departments are within the company, they have a pyramid- or matrix-shaped managerial organization according to the specificities of the departments and the size of the companies;

– the relations between the project manager, the directors of these departments and managers of the support services which operate through different areas of the project usually have a pyramid-shaped managerial structure led by company's general management;

– some support services, such as the quality service, assist the project manager in maintaining the overall coherence of the processor carrying out the innovation process.

The chain of events at the various steps of the creative research process can be described by the following figure:

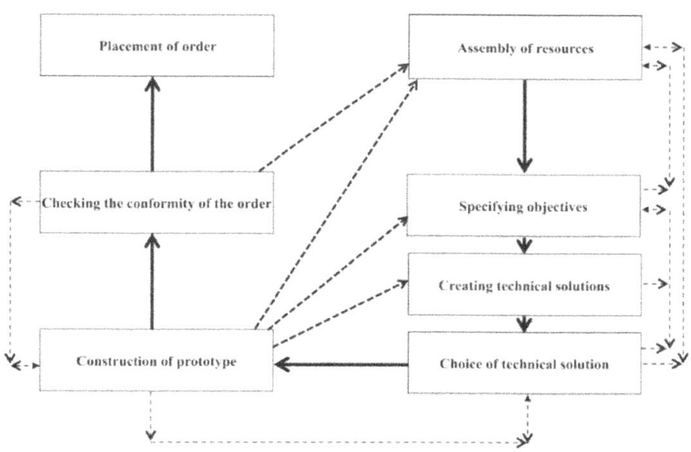

Figure 1.2. *Model of the steps of the creative research process*

Figure interpretation:

A new supplementary order can be created according to the conclusions gathered regarding the conformity of the creative research results to the initial order.

The assembly of resources necessary for the creative research to be carried out can occur at any time during the process depending on the results at the stage designed to specify the objectives to be met, the choice of technological or methodological solutions to be incorporated into a draft and the difficulties encountered in the construction of the draft.

The specific objectives of the creative research process can be redefined depending on the difficulties encountered in order to stop the technical or methodological solutions being developed, and according to the problems encountered to construct the draft of the solution identified.

These problems can also lead to research into new technical or methodological solutions or an examination of the technical or methodological solution being implemented.

Finally, checks carried out on whether the draft conforming to the original order can lead to the construction of a new draft with an alternative structure.

The results of one or several creative research programs can carry over to the materialization, whether physical or not, of a creation which can incorporate new functionalities within a product or service that the company could put on the market as part of its development policy or to reposition it on the market in relation to its competitors.

These results are often called "off-the-shelf" innovations. Their description is relatively accurate.

In general, the continuation of the innovation after the creative research phase is not systematic.

Innovation processes can be interrupted for a given amount of time (stop time), because the company management deems it unnecessary to immediately commit to new investments. The configuration stage of the innovation provided by the creative research program does not convince the company that there is a major momentary opportunity in continuing the development of the innovation to eventually include in the company's offer of services.

In order for the innovation activity to continue or restart, several factors need to come together:

– the innovation is based on one or some new concepts which are similar to other which have been successfully introduced in some products or services which do not respond to the same market needs; (Trend extension effect)

– the innovation initially seems to respond to a need identified through marketing, to which the market attaches a great amount of value and which is not yet being satisfied either by the company's current products and services, or by competitors; (Marketing breakthrough)

– by creating a break with the past, the innovation improves certain features of the company's product or service which are at the heart of the "company image"; (Image consolidation)

– the innovation substitutes a new technology for one or several others in a product which could eventually offer new performance levels for several functionalities or reductions of manufacturing, delivery or maintenance costs. (Technological breakthrough)

COMMENT 1.5.– It is possible for several of these "opportunity factors" to be present in a single innovation at the same time.

1.5.3. *Studies on the impact of a new innovation*

The configuration of an idea, resulting from creative research, however attractive it is for the team who contributed to its creation, requires impact studies to be carried out on the probable benefits or risks for the company if it takes the decision to use these results to design new products or services.

Above, it was noted that impact studies consist of the following stages:

– briefly characterizing products or services which could benefit from the innovation;

– identifying possible clients for these future products or services;

– characterizing future uses for products or services which would incorporate the innovation;

– conducting prospective studies on possible socioeconomic uses for the innovation, and its impacts on the company and its position;

– identifying future stakeholders for the innovation project;

– identifying stakeholders who will contribute to the development and production of the future products and services which will incorporate the innovation;

– conducting tests on the drafts of future products or services, at various steps in their production, on future clients and on all future stakeholders in the production system for the products and services.

One of the first parts of the study to be carried out examines the possible impact on the company's policy of including the innovation in an offer of services or products.

An innovation can cause a company to reconsider its strategic areas of activity. It may allow the company to access new market sectors by new offers, or to put forward a new offer which is significantly different from the previous one in the same market sector.

Every innovation also causes an evolution of the company's expertise. The nature of this impact regards the changes in qualifications and expertise of the staff, new technology to be introduced to the company, infrastructure changes to be undertaken, a redirecting of investments and so on.

In the same way, an innovation can modify the company's image or the offers in one of the strategic areas of activity. These changes to the image can represent opportunities to modify the company's policy.

Subsequently, it is necessary to very quickly establish the various stakeholders with which the company needs to develop new exchange relations in order to capitalize on the innovation and achieve the expected profits. These may be financial backers, providers, future consumers, future staff to recruit, etc.

It is also necessary to quickly envisage how these relations can be created for the company to make the profits which comply as best as possible with its policy. At this stage, it is simply a case of establishing the "staging" of these relations, in the theatrical sense of the term.

These scenarios help to identify some of the opportunity and risk factors which may arise in exchanges which the stakeholders targeted by the innovation.

This process can consist of the following stages:

– identifying all impact factors which could be considered as opportunities or risks that the innovation could produce either directly or indirectly in future relations with stakeholders, for the benefit or detriment of the company;

– specifying the nature of the benefits and risks, and providing an estimation of the related positive or negative predictions concerning returns on investments, by clearly showing the uncertainties in these predictions and the reasons why they occur;

– carrying out a sub-process of investigation, for each opportunity or risk factor, finalized by a description of their effects and an estimation of their probable contribution to the profits or losses for the company. Each sub-process has a predefined field of investigation. It must also show the main components in the future scenarios concerning the emergence of opportunity or risk.

The processors which carry out the sub-processes are made up of resources which belong to several departments (marketing, specialists in the new technology involved in the innovation, logistics, quality, etc.).

The draft tests on possible relations with stakeholders are essential in order to validate the relevance of these impact studies. The drafts can take very different forms depending on the nature of the relations to be established. It must be ensured that the information broadcast during these tests cannot be used by competitors.

1.5.4. *Feasibility studies in the innovation process*

A feasibility study is designed to check whether the use of the innovation by the company is realistic. It examines, in advance, how to assimilate the results of the creative research and the impact studies in order to design, mechanize and then provide the new products or services chosen to incorporate the innovation.

Above, it has been shown that feasibility studies include the following stages:

– conducting initial feasibility tests on the future products or services;

– identifying future stakeholders in the innovation project;

– conducting studies on the nature and feasibility of exchanges with future stakeholders in the project necessary for it to be fully realized;

– conducting performance tests on future products or services;

– conducting studies on the ability of the company to design production and distribution systems for the products and services which will incorporate the innovation by taking into account the configurations of the various definitive drafts of the innovation.

These stages are the most sensitive ones in the innovation process. The goal is to anticipate the directions of the processes of design, mechanization, construction, supply and use while taking into account opportunities, updated during impact studies. This anticipation can only be effective if it involves stakeholders who will contribute later in these various processes and who have experience in introducing an innovation into a new product or service.

The goal is also to reduce the risks which initially appear to limit the ambitions that the results of the creative research produced.

These feasibility studies can re-examine the results of the "creative research" so as to reduce the risks of failure in the use of the innovation, without losing out in terms of opportunities highlighted previously. It can also give rise to new opportunities.

When the company's management decides to do this type of feasibility study, it gives an order to the project manager, allocating to him the necessary resources and establishing more precise objectives than those which described each "innovation area" earlier on in this decision-making process. Deadlines are also imposed so that the feasibility study ends before the design launch of the new products or services which will incorporate the innovation.

The technical stages in a feasibility study are the following:

– the first sequence, before the design process starts, consists of anticipating the opportunities and risks associated with the stages of understanding, adopting and using the results of the creative research by those in the processor who are responsible for designing the new product or service;

– the second sequence consists of designing an initial supervision plan to implement during the stages of pre-study and study of the design process. It will be responsible for spotting new opportunities and capitalizing on them, but also for detecting new risks, and controlling their criticality;

– the third sequence consists of anticipating opportunities and risks associated with the stages regarding the design of construction processors and the delivery of the new product or service (mechanization of the product construction and the implementation of related services);

– the fourth sequence consists of designing a second supervision plan, to be implemented during the stages of mechanization, construction and delivery to the market of the new product or service. It is responsible for spotting the emergence of new opportunities and capitalizing on them, but also for detecting new risks, and controlling their criticality.

Supervision plans establish how added value is produced in the stages of design, mechanization, construction and delivery to the market of the future products or services which will incorporate the innovation, but also how exchanges will take place with stakeholders in the part of the company responsible for the various capitalization and innovation stages.

In summary, the feasibility study on the introduction of an innovation into new services or products is intended to anticipate risks of failures and of not keeping opportunities at bay, before their design and the plan for their construction and delivery.

It helps to clarify the decision of the company management on the use of the innovation, as described by the creative research results, in the design of new products or services.

It also gives future project managers, for the products or services, the systems which will be responsible for constructing, implementing and delivering them to clients, as well as methods for limiting the risks of not achieving their objectives, and capitalizing fully on opportunities to ensure optimal returns on investment.

The processors which carry out these studies are led by the project manager of the innovation process. He makes frequent use of other human resources from the design and mechanization processors. As a consequence, the human resources of these processors are mainly of variable structures and are made available to the project manager. At any moment, he may need the opinion of the collaborators participating in the creative research program.

This project manager leads the feasibility studies by also entering onto the "territory" of the design team in order to ensure optimal use of the added value produced by the innovation process.

He reports to the company management which requested the feasibility study and provides information to those responsible for the design and industrialization in the future.

1.5.5. *The decision-making stages of the innovation process*

As in any project, the decision-making stages are crucial. It is at this stage where the risks are the highest. A decision is always the choice to carry out an action by anticipating future events using the information available. The more that information is available which helps to predict the future, the more the risks of making mistakes are reduced. However, researching information carries a cost and takes time. Innovations cannot always wait. Commercial windows are becoming narrower and narrower to make an innovation profitable. Competition is always fierce, and methods of consumption and technological resources evolve extremely quickly. Decisions are therefore always taken with partial results. Decision makers must always be visionaries.

Our suggestion for the segmentation of decisions makes the task easier.

Decisions must always be taken during the project reviews.

As above, the decision-making process suggested is as follows:

– deciding to foster new ideas to help the progress of the company;

– deciding to initiate prospective studies on how to use the idea to benefit the company;

– deciding to guarantee the ability of the company to mobilize the resources necessary, internally and externally, in order to fully realize its innovation process;

– deciding contractual issues with these stakeholders, and building a processor to carry out the innovation process;

– deciding to progressively design drafts which give a more accurate representation of the idea;

– deciding to carry out prospective studies on the socioeconomic impacts of the future products and services which will incorporate the innovation;

– deciding to guarantee the ability of the company to adopt a production system for the future products and services;

– deciding to validate the innovation.

In an innovation process, going back to a decision is not a mistake. It is better to modify a decision when the situation requires rather than to persist with a mistake so as not to "lose face".

1.6. Classification of innovation processes

Innovation processes can be carried out in companies which are relatively predisposed, in terms of culture and time, to the specific mechanisms which are involved.

1.6.1. *Three types of company*

– Companies which are created through innovation (incubation); (E_1)

– Companies which periodically carry out innovation processes; (E_2)

– Companies which constantly innovate. (E_3)

The companies which belong to the first class have no company history and therefore have no company image, culture or infrastructure. The innovation process is the first process that they are undergoing, mainly relying on a "task force" managerial organization, led by the project promoter. These companies must seek out related support services which are now essential to ensure the survival or development of a company. The success of their innovation process is heavily dependent on the competencies of these external services.

The companies in the second class cannot implement a permanent department specifically dedicated to innovation. The process is often periodically integrated into design or mechanization processes which, due to their repetitive nature, are more standardized. It is carried out by a "task force" which frequently draws from its surroundings ideas which are well developed and often protected by patents. In other cases, the process is started by one person, or a small team, which releases a new idea which deviates from the company's typical practices. These companies have recognized specialized "expertise" which is the basis of their image. Creativity mainly relies on this shared expertise which helps to forge a common interest throughout the company. The innovation must be prepared to expand this expertise and must contribute to strengthening their image. These restrictions limit the capacity of an innovation process to start afresh except in the case of a new trend in society (a fashion, for example) which naturally highlights the company's expertise and automatically encourages it to undergo an innovation process. There are consequently high chances of success and the risks are relatively low when the company has a culture of "project management".

The companies belonging to the third class have an organization which is specifically dedicated to innovation. This organization tends to become burdened by the weight of feedback and its efficiency is limited by the lack of spontaneity in its creativity. They are able to produce new levels of expertise or images by dividing up the new activities created by an innovation process which can open significant and sustainable new markets. These new markets allow these companies to guarantee their future by aligning their developments with important social trends leading to economic development.

1.6.2. *Types of innovations*

Innovations are of the following types:

1) the creation of a new product or service which responds to a need which is as yet unanswered; (Type I)

2) the introduction of new functionalities into an existing product or service which allows it to better respond to a need which is only partially satisfied; (Type II)

3) the introduction of new technology into a product which improves the performance of previous functionalities; (Type III)

4) the introduction of new practices into an existing service which improves the performance of previous functionalities. (Type IV)

1.6.3. *Correlations between types of innovating companies and types of innovation*

Experience shows that there are correlations between the types of innovating companies and types of innovation. They are shown in the table below:

	I	II	III	IV
E_1	x			
E_2		x	x	x
E_3	x	x	x	x

Table 1.1. *Correlations between types of company and types of innovation*

From now on we will only distinguish between two types of situation. The first includes the innovations which create new products or services (creation). The second includes the innovations which lead in some way to an evolution of existing products or services (modification).

EXAMPLE 1.2.–

– the invention of the steam train, GPS and prosthetic hip replacements are type I innovations;

– the creation of the mobile phone and related services, remote cardiology and bank card payment are all type II innovations;

– the electrification of the rail network, disk brakes on cars and the introduction of ceramic prosthetic hip replacements are all type III innovations;

– fast food, attending doctors and the sub-contracting of postal services in rural areas are all type IV innovations.

The distinction between the different types of innovation is important for several reasons:

– there are different triggers for the different types of innovation;

– the processes which carry out the innovation projects consist of different stages;

– the risks connected to the various innovations differ;

– the stakeholders in the processes carrying out the innovation projects do not have the same profiles;

– the impacts on the changes to behavior, attitudes and consumption methods are different;

– the impacts on the organization of the consumption process are also different.

1.6.4. *The specificities of type I innovations*

1.6.4.1. *The triggers*

For type I innovations, the triggers are "creators" able to identify needs which have not yet been answered in a population ready to expend a lot of energy in order to satisfy it, and connect them with an idea expressed by the draft of an original system configuration which could give a credible response to these needs partially or in full.

The creator is often in direct contact with the population which is experiencing the need or needs. It is even frequently a member of this population. Its research is guided by the awareness of the significance of such a need or needs. For instance, type I innovations in medicine are generally triggered by doctors who know technicians or engineers. The doctor is the "idea promoter" and the engineers of the company are the "developers" of the idea. They form a pairing which triggers the innovation process.

It can be observed that this type of innovation is often triggered by a pairing made up of "idea promoters" and "developers". The director of a research and development department for a large consortium, which has registered more patents than any other in France in the last few decades, often noted that the innovation processes were always triggered within his company by what he called "shock pairings". This refers to an idea promoter and a developer who are known within the company as often being in conflict and to whom a targeted innovation mission is entrusted regarding a need identified by marketing.

We often heard other research and development directors saying that the best innovations within their companies were often triggered by a group of enthusiastic individuals who carried out the creative research in private, alongside their official professional activity and with few resources.

A type I innovation can give rise to a new company. Today, there are structures which facilitate the creation of this type of company ("incubators").

1.6.4.2. *The processes*

The creative research stage of a process producing a type I innovation is fundamental. It consists of several disconnected sequences interspersed with "dead time".

Each sequence is carried out by a specific production system which can differ from one sequence to the next. The output data of a sequence can differ from the input data of the following sequence.

These processes are very long and highly unpredictable. Several innovations can be developed from each sequence. There are high risks of failure.

A company cannot take on all the sequences and necessary funding by itself. Generally, public authorities participate in financing this type of operation. Research must be carried out alongside. This often requires several scientific problems to be overcome in order to access more fundamental research.

Only when there is a convergence between the results of fundamental and applied research can a type I innovation emerge.

There are high risks linked to the decision to design a product or service which embodies a type I innovation. This cannot be taken without in-depth impact and feasibility studies. The system configuration results from the different sequences of creative research, which are difficult to predict and change from one sequence to the next. It is therefore not desirable to start these studies too early on. Moreover, the variety and quantity of information to be gathered and processed in these studies means that they cannot be carried out simultaneously, but one after the other, because the results of the first can significantly influence those of the second.

1.6.4.3. *The nature of the risks*

The risks linked to type I innovations are highly unpredictable because of the significant amount of new ground being covered by such innovations. Risks are therefore permanently anticipated throughout the various sequences in the creative research process, and impact and feasibility studies. They can be risks of any kind.

The anticipation or even the prevention of risks is extremely difficult. Therefore, controlling risks depends largely not on restrictions to creativity but on observing and correcting as quickly as possible intermediate results that could produce final results which do not comply with the expectations guiding the decision taken at the launch of the innovation process.

1.6.4.4. *The stakeholders who produce the innovation*

The stakeholders who participate in projects for type I innovations belong to inter-disciplinary teams which combine researchers, developers and well-informed consumers of products or services with similar functionalities to the future creation. These stakeholders belong to different structures which have decided to pool their skills and resources to carry out this type of innovative projects.

Today, these initiatives are promoted by the creation of competitiveness clusters.

1.6.4.5. *The impacts on consumption methods*

In general, type I innovations have impacts on the development of land, social evolutions and some aspects of individual behavior of a large sector of the population. Their entry into consumption systems is gradual, and requires the creation of a new environment adapted to the use of the innovation which must be envisaged and designed at the same time as the innovation.

For the innovation to be well received, it is necessary to carry out work on all areas of the consumption system so that the innovation is able to respond to the essential need of future consumers.

This type of innovation must consider "attractive" functions, but also "repelling" functions especially connected with changing habits, or even "unintended" functions which cause disturbance, for instance, to other methods of consumption or lifestyles.

For instance, the emergence of large superstores, as opposed to small shops, gradually changed the organization of service to the individual. Changes in consumption methods were gradual, with significant changes in land development, and "self-service" considered, by some consumers, as repelling factors. This innovation led to the disappearance of some social links between shopkeepers and customers, but also between the customers themselves.

1.6.5. *Specificities of type II innovations*

1.6.5.1. *The triggers*

For type II innovations, the role of circumstances is a little different. They are often created by a company which already exists and has a marketing department able to identify a need which is only partially satisfied by the company's products or services, or those of competitors. Moreover, satisfying this need will lead to a strengthening of the company's image by giving rise to profitable areas of originality compared to the competition.

The company then assembles its creators to research solutions, firstly functional and then technical, by fully capitalizing on the expertise of the company, or those with similar practices and culture.

In the agro-food sector, moving from a high-fat dairy-based product to a lighter version while keeping the same flavors is a type II innovation.

This innovation is only of interest if the company has or wishes to have an image based on the production of healthy foodstuffs. In this case, it must have a structure which is able to permanently carry out innovation processes with this strategic aim so that in the long term the market naturally associates the company name or some of its brands with the idea of "food which maintains or improves health".

Generally speaking, type II innovations are only profitable if they align with a long-term strategy for strengthening the company's image or for creating a new image. The company must therefore have a permanent structure which constantly undertakes innovation processes in compliance with the defined strategic direction, and which gradually builds a network of relationships with specialized providers especially with certain research centers.

1.6.5.2. *The processes*

The processes which produce type II innovations are extremely varied. They generally depend on the size of the innovation(s) involved in the project.

Firstly, it is necessary to distinguish between single- and multi-innovation processes.

For instance, a project focusing on the car of the future is a multi-innovation type II project. The first GPS integrated into a car was also a type II innovation. The combination of a medicine with an innovative device which transports the medicine to its target to treat an illness is another type II innovation.

The processes which carry out this type of innovation are characterized by the fact that they use existing technology and they adapt it to integrate it into an existing product in order to enhance the functionalities of a product to better satisfy a major need of the market.

These processes are fully carried out by the company.

The risks of not achieving the expected performance of the new functionalities remain, as do the risks of not satisfying the needs as the market would have liked. Even more seriously, the company can also make a mistake in terms of the importance that the market attaches to the response provided to this need. It is also possible for more unexpected situations to arise: the new functionalities give rise to new needs which obliterate the interest in satisfying previous needs.

We will also see later that the introduction of new functionalities intended to be attractive to the market also introduces new repellent functions which prevent a successful entry into the market. In this case, the innovation is a commercial failure.

The introduction of a quality process in this type of innovation process is particularly essential for ensuring, above all, a better containment of the various risks.

The creative research stage of the innovation process does not generally require supplementary scientific research. The technologies or methodologies are known, but have not yet been used to perform these expected functionalities.

However, creative research must focus on the significant and various restrictions imposed on the project which are due to the fact that the innovation needs to be integrated into an existing product or service. It also often needs to be integrated into an operating system of a product or service which has become established, and operating habits which need to be taken into account. This collection of restrictions is a limiting factor to the creative research stage and must be taken into account as early as possible in order to avoid modifications during this stage of the innovation process, which often entail significant consequences on managing risks, costs and deadlines.

These observations highlight the need to carry out feasibility studies in parallel with creative research by scheduling joint project reviews for these two stages in the process.

The feasibility study process can be more standardized since it is reproduced almost identically with one introduced to the next within the same company. It is possible to improve its performance by capitalizing on feedback produced.

Impact studies must start before the creative research and continue alongside it until the results of the creative research are configured. It is only with the final configuration of the innovation, deemed feasible and compatible with all the restrictions imposed by the existing configuration, that it is useful to research all the opportunities to grasp to optimize the value of the innovation.

1.6.5.3. *The nature of the risks*

The risks attached to type II innovations can be more easily anticipated and located. However, it is important to remember that a product is a complex system. Any integration of a new functionality can have direct or indirect effects on other functionalities even if they are not seemingly connected. In the same way, in order to grow the functionalities of a service, malfunctions can be caused to another aspect of the process by creating cause–effect links between certain actions contributing to its production, without this being clearly visible. In general, this type of innovation is created in companies which have decided to focus on innovation to build an image to differentiate themselves from their competitors. In this case, innovations occur frequently. By using feedback, it is possible to locate the main areas of risk and anticipate them.

1.6.5.4. *The stakeholders who produce the innovation*

The stakeholders participating in projects focusing on type II innovations are generally engineers or technicians from the same company who belong to a department specialized in the production of this type of innovation. Their source of inspiration may come from other workers in the company or externally by practicing “economic intelligence”. The marketing services generally participate in these projects to identify the commercial target and to provide information on consumption methods of the target markets and the state of the competition.

1.6.5.5. *Impacts on consumption methods*

Type II innovations have less impact on social structure than type I innovations. However, in general, they have a significant impact on a limited and targeted sector of the population of future consumers (market sector). They can cause a significant change in certain professions by modifying practices and enhancing some activities which rely on the innovations integrated into the products or services and which are central to the profession.

For instance, remote cardiology modifies the organization of care and strengthens the role of some staff members who assist cardiologists. The cardiologists themselves are more involved in the prevention of cardiovascular diseases than in the past.

The introduction of in-built electronics in cars caused changes in terms of diagnoses, requiring significant investments from mechanics, and led to an evolution in the knowledge and expertise of repair shop staff.

1.6.6. *Specificities of type II and IV innovations*

1.6.6.1. *The triggers*

For type III and IV innovations, functionalities are already characterized and the innovation is limited to researching technology or new practices for the company to use on the product or service, which may have already been tested in other contexts. Therefore, the triggers are usually technicians or engineers responsible for "economic intelligence" within the company.

In the competitive context which we find ourselves in today, every company must constantly research type II and IV innovations, in order to guarantee its future. It can draw sources of inspiration from the dynamics of global technological and social changes.

These innovations, which can generate significant returns on investment, do not entail great risks for the company. They are limited to identifying, among its non-competitive environment, the use of new technologies and methodologies and to studying their transfer into the field of application of the functionalities of the products or services that the company wishes to improve in order to strengthen its competitiveness.

The trigger of these innovations is the combination of a clear need to evolve the performance of certain functionalities of the company's products or services, and the discovery of "off-the-shelf" solutions by observing dynamics of evolutions in the company's surroundings.

The benchmark can be an excellent trigger for this type of innovation.

1.6.6.2. *The processes*

The processes for type III and IV are identical. The functionalities of the product or service are unchanged, but the objective of the innovation is to significantly increase the performance of certain functionalities by introducing new technologies and methodologies.

Creative research is limited, as before, to identifying existing technologies or methodologies by taking advantage of feedback regarding their use in other contexts, and to introducing them into a new configuration of the product or service to improve certain aspects of their performance.

As a result, the objectives of this research are well defined right from the start of this process. Uncertainties regard the performance of these new technologies in a new setting, the impacts on how to carry out a service and to ensure it is appreciated by beneficiaries.

The creation is limited to adjustments which are based more on a perfect knowledge of a technology or methodology than on imagination. Specialist expertise plays a key role in carrying out this type of process.

The impact study is essential in this type of innovation. It allows the project team to identify, as best as possible, the expertise created by capitalizing on feedback on the introduction of technology or methodology in other circumstances. They can then attempt to use similar situations to save time and better contain risks.

This study must also ensure that the new configurations of products or services are well perceived by future consumers as having real innovations which justify their attraction. Indeed, technological or methodological transpositions can be seen as simply gadget add-ons without offering any new usage value.

The feasibility study mainly examines how to introduce new practices which are based on different cultures.

1.6.6.3. *The nature of the risks*

The risks attached to type III and IV innovations are identifiable and, in general, are treated in the same way as those which are produced by type II innovations. It is common for an innovative project to combine type II, III and IV innovations.

1.6.6.4. *Stakeholders who produce the innovation*

The stakeholders who participate in the projects for type III and IV innovations are workers within the same company or workers for the company's providers. The directors increasingly encourage their providers to co-produce this type of targeted innovation with them, using a set of functional specifications. Major new risks can emerge at this stage. These are due to a lack of experts in the project team specialized in the new technology introduced into the innovation. It is the provider who has the expertise rather than the company. The company must ensure that it keeps hold of this type of expertise in order to maintain the leadership role in its relations with the providers. The company can also assemble its employees on a wider scale so that they become the drivers of proposals of this type of innovation (participative innovation).

1.6.6.5. *Impacts on methods of consumption*

Type III innovations have a weaker impact on consumption methods. They mainly affect the way of delivering product maintenance and after-sales services. They make products more robust and focus consumers' attention on the uses that they can make of these products – they have fewer concerns over the way the products they use work. The performance of the products' technical functions then becomes obligatory. Any drop in performance of these functions cannot be tolerated.

Type IV innovations are often intended to increase the consumer's comfort or passiveness when they co-produce a service with a contractor. When they affect services associated with products, they facilitate the use of the product.

1.7. Conclusion

The information developed in this chapter confirms the particular complexity and variety of innovation processes.

Every innovation process is a complicated project aiming for radical change. It is a "one-shot" process which can rely to a greater or lesser extent on feedback as a guide. The value of the innovation is closely linked to the significance of the risks and opportunities which emerge during the process. Containing these risks is not the same as "avoiding the risks", but identifying them relatively quickly, accepting to take them regardless of their initial criticality and gradually attempting to reduce the criticality to acceptable levels. The permanent opposition of "needs", "creation" and "production", based on three very different logics, is absolutely necessary.

The integration of a competitive quality process into the innovation process is therefore a complex action which must take into account all the specific characteristics of the process and the processor carrying it out. It is necessary to examine all these specificities in order to design the individual configuration of the quality process to be integrated into the innovation process.

Although these specificities entail different methodological implementations from one process to the next, in order to obtain the desired level of quality, it should be noted that they are based on the same logical principles. It is these principles that we will look at to identify and characterize in the next chapters.

2

Competitive Quality of an Innovation

2.1. Introduction

The objective of this chapter is to review the general principles of competitive quality processes which can be integrated into innovation processes to increase their chances of success in a highly competitive context. These characteristics are discussed more generally in our book entitled *Competitive Quality Strategies* [MAI 13].

Today, most companies already find it difficult to implement these principles in production processes, especially when they carry out service activities, because they are significantly different from traditional quality processes which have been applied for a number of decades.

These problems become greater when they affect innovation processes which must remain extremely flexible with inevitably high levels of uncertainty, and require real-time improvisation.

The creation of an innovation naturally focuses most of the energy expended by its stakeholders on gradually creating the innovation's configuration. This is fundamental to the production of added value.

The other activities of the project team are considered as supplementary and often as restrictions, or even annoyances. "If we could replace the members in the market targeted by the innovation, and get rid of the other stakeholders in the innovation process, and just aim to please the members of the project team, it would be ideal". This message is often implicitly or explicitly conveyed by the creators. Some even believe that this practice of creation is necessary in order to produce value.

Creation is an extremely selfish process.

In Chapter 1, we discussed how an innovation process cannot be limited to this type of creation. It can only succeed if the project team producing the innovation is able to obtain the necessary resources from its surroundings and if it is able to make the project attractive to stakeholders who will contribute to its success.

We should recall that:

> *An innovation process is the gradual materialization of new ideas guided by the projection of a very uncertain future, in terms of its methods of use and the returns on investment that they will produce.*

Since they are complex and are carried out in increasingly highly competitive contexts, it has become essential to introduce a competitive quality process in these innovation processes.

The pressure of competition, the acceleration of change, the increasing social constraints and the reduced time in which a product or service retains its attractiveness all indicate that there is an ever-lower chance of an innovation to succeed.

By integrating a competitive quality process *in the production of exchanges which need to take place between stakeholders in an innovation process, the company increases its chances of reaping the benefits which live up to its expectations.*

The production of quality and the creativity are the two main sources of added value in an innovation process. The role of the team responsible for the innovation project is to use these sources in the best way possible so that the project generates sustainable exchanges, which are beneficial for the company, with stakeholders participating in its success.

2.2. The concepts which characterize competitive quality

Generally speaking, a quality process in a company aims:

> *To allow the company to benefit, in as sustainable a way as possible, from predicted exchanges with stakeholders by prompting certain judgments they have with regard to the quality of their activities.*

Stakeholders are socio-economic participants who exchange products or services with the company.

The meaning of "in as sustainable a way as possible" can be interpreted in a different way depending on the company and each class of stakeholder, according above all to their policy.

Henceforth, these particular judgments will be referred to as: "competitive quality judgments".

The complexity of this type of quality process means that its effectiveness is constantly called into question.

We will see that the pieces of information released by a company to stimulate competitive quality judgments from stakeholders are very different in nature. This information may concern a product's technical performance, its aesthetic aspects, a piece of documentation, any sign of confidence in the company's ability to deliver what it is announcing, the attitude of some members of the company who are in contact with the stakeholders, the accessibility of a service and so on.

To facilitate the production of an innovation's quality, it is desirable to organize its stakeholders in a different order than the one which has been proposed previously using the notion of "strategic areas of activity". This change is justified by the "process" approach in modeling this production which gives a fundamental role to time in its organization.

Previously, we distinguished two strategic areas of activity: one which is associated with the innovation processor and the other with the future business unit (BU) which will capitalize on the innovation.

The first strategic area is made up of:

– a package of services that the processor offers to members of the company's BU which will benefit economically from the innovation (project offer);

– the innovation processor itself (production system for the innovation process);

– the strategic area of activity of the BU which arranges the clients of the innovation processor;

– the family of stakeholders who produce the package of services (project team);

– the family of stakeholders, external to the project team, who provide certain components of the service package, or resources which are then processed by the project team to produce these services (project providers);

– the family of stakeholders who invest in the project hoping to benefit economically from the innovation (project investors);

– a collection of parties connected to any of the above stakeholders and which have an indirect effect on the project.

The second strategic area is made up of:

– the service package which the BU will offer to its clients by utilizing the innovation (supply);

– the processor which will produce the package of services (BU);

– the family of stakeholders who will make up the future clients of the service package (potential market for the innovation);

– the family of stakeholders who will produce certain components of the service package in the "BU" (workers);

– the family of stakeholders external to the "Business Unit" who will provide certain components for the service package, or resources which will then be processed by the company in order to produce these services (providers);

– the family of stakeholders who will invest in the BU hoping to benefit from the production of the services (investors);

– a collection of related stakeholders who are connected to any of the above stakeholders;

When we address quality processes integrated into innovation processes, we are firstly focused on the family of stakeholders who will constitute the clients of the future service package (innovation's potential market).

This class of stakeholders will be marked as (C).

We will then consider:

– the family of stakeholders who produce the service package (project team);

– the family of stakeholders, external to the project team, who provide certain components of the service package, or resources which are then processed by the project team to produce these services (project providers);

– the family of stakeholders who invest in the project hoping to benefit economically from the innovation (project investors).

These stakeholders will be marked as (I).

Finally, we will address exchanges which take place with:

– the family of stakeholders who will produce certain components of the service package in the "BU" (workers);

– the family of stakeholders external to the "BU" who will provide certain components for the service package, or resources which will then be processed by the company in order to produce these services (providers);

– the family of stakeholders who will invest in the BU hoping to benefit from the production of the services (investors).

These stakeholders will be marked as (P).

There are a large number of stakeholders, but they do not all have the same level of importance, and actions on their competitive quality judgments are not planned for the same time or with the same intensity in the innovation process. This intensity depends on the level of competition encountered during the planned exchanges with them.

In general, an innovation is limited to a product or service designed for the class of stakeholders who will use it. It is difficult, even pointless or risky, to innovate simultaneously in all directions.

Relations with other classes of stakeholders remain more conventional. This does not mean, however, that it is not necessary to introduce a competitive quality process in these interactions because it is becoming increasingly likely that the project will be met with stiff competition.

For instance, in relations with investors participating in project funding, the pieces of information offered by the company are traditional and often standardized. These are business plans, proof of the company's ability to successfully complete the project, results of market research and so on.

An economic entity can belong to more than one class of stakeholder. The future beneficiaries of the products or services who will include the innovation can also be the co-producers of the innovation process, investors, providers or social regulators.

For instance, an innovation process in the medical sector often draws from doctors who will be future users of the innovation. This situation is often the source of inevitable conflicts of interest when these doctors also play the role of social regulators because they have such valuable expertise.

In some start-ups, these doctors are also occasionally investors and providers.

An innovation process must therefore not simply concern itself with the quality of the innovation's production, but also with the quality of the other services offered to this group of stakeholders.

For example, it is necessary to attempt to attract investors looking for returns from the project by providing them with attractive economic projections, but also at acceptable levels of risk. Members of the project team need to be encouraged to participate in the project by accepting the obligatory specifications to fulfill. To do this, it is necessary to stimulate perceptions of quality that the other activities in the company do not do with as much power. Future producers need to be provided with products or services which incorporate the innovation, a set of specifications which is credible and attractive, showing how the innovation will enhance the future products or services, and how they will be produced. The department responsible for the protection of the environment, persons and property must be provided with evidence showing that the future products or services respect current regulations, within acceptable risk levels.

This model simply shows the situations which are often managed in the most informal way in every innovation process. It simply suggests extending the field of "competitive quality" into a larger base of stakeholders.

In a start-up, an important nucleus of stakeholders simultaneously belongs to several stakeholder groups. The project promoter may be a future worker, an investor or even occasionally a provider. A member of the project team can be an investor, a provider or a future client. In this way, specific complex links are forged with the company. It is preferable for these links to be formalized in contracts as soon as possible.

The project manager leading the innovation process can be assisted by numerous support services to carry out the competitive quality process integrated into the project.

The various support services in each class of stakeholders are respectively as follows:

– financing service for investors;

– marketing service for clients;

– purchasing service for providers;

– human resources services for workers;

– services responsible for relations with social regulators (environment, safety, legal, etc.);

– quality service.

Once again, depending on the size of the company, these various services can be carried out by the same team, or even the same person.

In the following, we will see that the quality service plays a role of *global dynamic regulation* of the quality process, whose goals are to:

– encourage the most profitable use of stakeholders' existing competitive quality judgments during the innovation process;

– ensure the design and maintenance of systems to regulate the product of information released by the innovation process which stimulates stakeholders' competitive quality judgments;

– contribute to the attractiveness of and confidence in the company, regarding its ability to complete its innovation projects successfully.

COMMENT 2.1.– A project promoter looking to found a company using an idea which could trigger an innovation process can find these support structures in an incubator.

To define the meaning of the concept of *competitive quality judgments*, and how they are produced, it is useful to introduce the concept of perceived quality in the following way:

> *The perceived quality of an innovation process by one of its stakeholders is the end result of a collection of perceptions to responses offered by the production process to its expectations.*

In general, the main classes of expectations which lead to these perceptions can be described in the following way:

– to satisfy needs by the use of the production process, its results or their effects;

– to have confidence in the ability of the process to comply with the objectives stated;

– to understand, obtain, adjust and use this production, its results, and their effects;

– to make the work carried out to understand, obtain, adjust and use them profitable.

Box 2.1. *"S.C.A.R." expectations*

These expectations are known as "scar" expectations because they are made up of the Satisfaction of needs, a search for Confidence, Accessibility of the product or service offered, and Rendering work required to obtain and use it profitable (S.C.A.R.) [MAI 11].

This classification of expectations leads to a similar classification of the main perceptions which allow the stakeholder to reach a judgment on the perceived quality of the innovation process, before and at the moment it commits to participating.

Four classes of competitive quality perceptions can be distinguished. They are described in the box below.

The main classes of competitive quality perceptions of an innovation process can be expressed as follows:

– Perceptions of expectancy regarding:

- the satisfaction of expectations by certain components of the innovation production process, its results or the effects it causes;

- the ability to use these components;

- the ability to make the work required to understand, obtain, adjust and use them profitable.

– Perceptions of confidence in the ability of the innovation process to comply with the objectives given.

Box 2.2. *The classes of competitive quality perceptions of an innovation process*

These perceptions start before an implicit or explicit contract is agreed upon between the stakeholder and the project. This continues during the exchanges between the two entities, which leads to the innovation process.

It is at the start of this stage where exchange relations between the project and the stakeholder begin or halt. It is at this moment when perceived quality plays the most significant economic role. It can be an initial exchange, or a renewal of a previous exchange. In the latter case, these perceptions take into account previous exchanges. They have an influence on perceptions of expectancy, and confidence at stake at the moment of contract renewal. Consequently, perceived quality in previous relations is a key component to be taken into account in the production of quality which is necessary at the renewal of exchange relations. This becomes even more relevant if the company is subject to high levels of pressure from competitors.

In the event when an innovation is integrated into a product or service, previous exchange relations with a stakeholder inevitably influence its perceived quality in the innovation process. It is difficult for an innovation to cancel out a bad reputation of the company that may exist.

It is necessary to differentiate between perceived quality and compliance quality.

DEFINITION 2.1.– *The compliance quality of an innovation process is the compliance of its production, results and effects that it produces with the specification established as objectives.*

These objectives, established by the process leader, do not systematically lead to competitive quality perceptions which allow the company to reap the benefits expected.

Each stakeholder makes a value judgment about its competitive quality perceptions. This judgment plays a key role in allowing the company to capitalize fully on the exchanges with the stakeholder.

DEFINITION 2.2.– *A competitive quality judgment on an innovation process made by a stakeholder comes from an overall appreciation of their competitive quality perceptions, during the exchanges which take place as the process advances, allowing the innovation project to benefit from support given by the stakeholder who satisfies the expectations of the promoter, despite the pressure of competition.*

The relational dynamics which are activated by quality in an innovation process between a company and its shareholders can be modeled in the way shown in Figure 2.1.

This is a particular type of economic process known as "win-win" for the innovation project and its stakeholders.

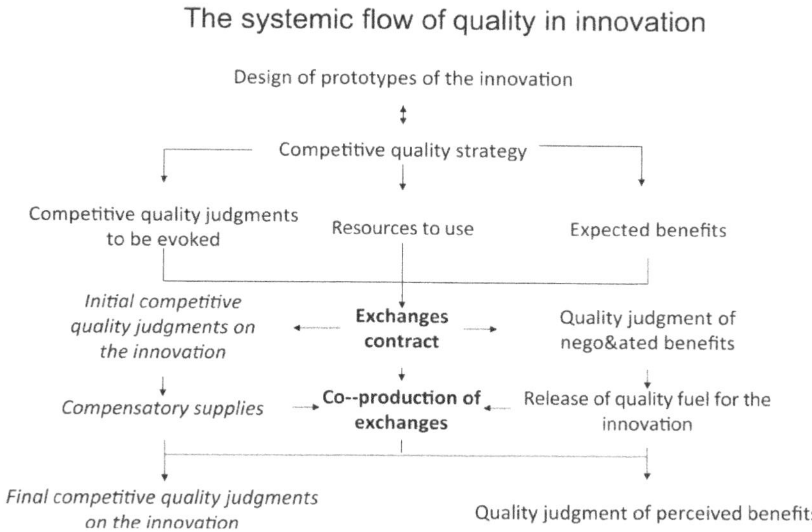

Figure 2.1. *The dynamic flow of quality in an innovation process*

In the figure, we have used terms which can be defined as follows.

DEFINITION 2.3.– *The integration of quality into an innovation process is also known as the "quality fuel" of the innovation. It consists of a collection of information, released by the innovation process, its results and the effects that they produce, which have an impact on the stakeholders' competitive quality perceptions.*

COMMENT 2.2.– The term "competitive" can be added to "quality fuel" when it helps to prompt the stakeholders to make competitive quality judgments on the innovation process.

The final competitive quality judgment made by a stakeholder on the innovation process is the quality judgment which it makes at the moment when an exchange relationship with the company comes to an end. This judgment will influence the decision that it will need to make regarding possible collaboration with the company in a new innovation project.

This quality integration starts as soon as the project promoter designs the first prototype of the innovation.

The first task is to design what is known as a "competitive quality strategy".

This anticipation stage is essential. It will allow the company to capitalize fully on the innovation, to strengthen its position against the competition in the socio-economic environment.

A competitive quality strategy regarding exchanges with a stakeholder:

– describes the main characteristics of the competitive quality judgments on the product or service offered to the stakeholders, to encourage them to provide the resources expected by the project leader, in the most profitable way possible;

– defines the main characteristics of the compensatory resources which allow the project to reach its objectives, ensuring sustainable contributions from the stakeholder;

– chooses the most profitable processes and resources to implement within the project in order to prompt competitive quality judgments and benefit from the compensatory resources.

A quality strategy supposes that the project leader has already defined the support that he hopes to obtain from the stakeholder.

After drawing up a competitive quality strategy, the project leader produces and sends announcements to the stakeholder describing certain characteristics of the innovation process, its results, and the effects it will produce. These announcements make up part of the *competitive "quality fuel"*. They prompt an initial competitive quality judgment on the innovation encouraging the stakeholder to engage in an exchange relationship with the project.

After drafting a contract with the stakeholder, the project leader makes a quality judgment on the expected benefits of the project which have been negotiated with the stakeholder. This can be a competitive quality judgment, when the company has the choice between several stakeholders.

As the innovation process progresses, it releases pieces of information which are received by the stakeholder. This prompts competitive quality perceptions, which can lead to an evolution in its competitive quality judgment, and supports or opposes the contributions that it has committed to providing once the contract has been agreed upon with the project leader.

The summary of these perceptions produces the stakeholder's final competitive quality judgment.

In a symmetrical way, the project leader summarizes the quality perceptions of the stakeholder's contributions. This summary produces a final quality judgment of the benefits which result from the exchanges with the stakeholder.

2.3. The use of competitive quality in an innovation process

An innovation process is a catalyst for placing the company which carries it out into a network of economic exchanges allowing it to reap the benefits to ensure its future through policy choices.

Competitive quality is used in an innovation process to attempt to prompt competitive quality judgments from its stakeholders, thus leading to their contributions helping the success of the project in the most profitable way possible.

We can define this general formula as follows.

> A competitive quality process in an innovation process is intended to ensure that the stakeholders' competitive quality judgments contribute, in the most profitable way possible, to:
>
> – creating a purchasing value for products or services which integrate the innovation in order to respond, better than oppose, to the needs of the market which the company wants to occupy in accordance with its policy;
>
> – ensuring the benefits expected by the company as a result of these new products or services;
>
> – facilitating the acquisition of resources necessary for the product to produce the expected purchasing value;
>
> – respecting the deadlines imposed for entering the product or service onto the market in compliance with commercial demands;
>
> – forging and sustainably strengthening the desired links between the company and the target market due to this offer of products and services;
>
> – respecting the regulations and safety rules of the countries where the market is located;
>
> – contributing to enhancing the image of the products or services which incorporate the innovation or the company's image.

Box 2.3. *The objective of a competitive quality process in an innovation process*

This objective is complex because there are many stakeholders, because quality judgments are often irrational, and evolve significantly, because competitors constantly attempt to reduce their attractiveness and because it is difficult to anticipate long-term returns on investment.

This objective becomes even more difficult when the production of an innovation process' value is subject to constant influences which change its direction, and when the resources provided by some stakeholders contribute to the attractiveness of the products or services provided by the project to others. These permanent interactions between perceived quality of certain products or services offered to stakeholders and resources provided for the innovation process to produce the perceived quality of other products or services sometimes appear to give rise to a "chicken-and-egg" situation. To avoid these vicious circles, a time-based order must be imposed in the execution of these actions to produce competitive quality in the products or services offered to stakeholders.

It is also necessary to adapt the way in which this competitive quality is produced for each of the five main areas of any innovation process, which are:

- creative research;
- impact studies;
- feasibility studies;
- organizational stages;
- decision-making stages.

2.4. A model of the competitive quality process in an innovation process

To help in mastering this complexity, we suggest a number of models which serve as logical references.

The generic model which we propose to represent a competitive quality process integrated into an innovation process can be defined using the following steps.

A "competitive quality process" integrated into an innovation process includes all of those specific actions in the process which:

– target how best to use the competitive quality judgments of its stakeholders to obtain the performance expected of the innovation;

– design specifications for the competitive quality fuel which the process must create during exchanges of products or services with stakeholders, in order to produce competitive quality perceptions maximizing the chances of reaching the desired targets;

– ensure, at the lowest cost, the compliance of the competitive quality fuel created with their specifications;

– evaluate the predicted effects or those actually produced by the creation of this quality fuel on the performance of the innovation;

– envisage and implement possible improvements regarding: the design of competitive quality fuel specifications in the innovation process, the management of their compliance with these specifications and their effects on company profits produced by the innovation;

– capitalize on feedback.

Box 2.4. *The main functions of a competitive quality process integrated into an innovation process*

These actions can be grouped into classes which can be identified according to their contributions to technical functions which can be described as follows.

The technical functions of a competitive quality process integrated into an innovation process can be described as follows:

– define a competitive quality strategy in exchanges with stakeholders in the innovation process;

– for each class of stakeholders, design scenarios regarding production and release of competitive quality fuel, competitive quality judgments and contributions from stakeholders who have the best chance of fulfilling the competitive quality strategy (marketing quality);

– design specifications for competitive quality fuel offered to stakeholders, so that scenarios have the best chance of being fulfilled with the minimum amount of energy expended (design of competitive quality fuel);

– design emitters of competitive quality fuel integrated into the production systems of products or services offered to stakeholders, capable of systematically complying with the competitive quality fuel specifications (design competitive quality fuel emitters);

– when each competitive quality fuel emitter is activated, ensure that it is able to systematically comply with these specifications (qualification of competitive quality fuel emitters);

– check the compliance of the various processes related to marketing quality to their scenario (validation of marketing quality);

– ensure, during the production process of competitive quality fuel, that the risks of non-compliance to their specification maintain an acceptable level of criticality (supervision of the capability of competitive quality fuel emitters);

– eliminate areas of non-compliance with specification as soon as possible in the competitive quality fuel production process (processing of non-compliance with specifications of competitive quality fuel);

– oversee the impacts of competitive quality fuel in the innovation process on the competitive quality perceptions of various stakeholders (supervision of perceived quality);

– ensure that the stakeholders' competitive quality perceptions lead to the supply of compensatory resources in compliance with the expectation of the competitive quality strategy (ensuring expected benefits);

– track feedback in order to improve the efficiency of competitive quality processes which will be integrated into new innovation processes (improving the ability of the company to use stakeholders' competitive quality judgments on an innovation process).

Box 2.5. *The technical functions of a process to ensure competitive quality*

	Targets	Design	Ensuring compliance	Performance evaluation	Improvement	Capitalization
Creating a competitive quality strategies	X				X	X
Creating marketing quality	x				x	x
Designing specifications for competitive quality fuel	x	x			x	x
Designing specifications for emitters	x	x			x	x
Qualifying emitters			X	x	x	x
Validate marketing quality				x	x	x
Supervision of emitters' capability			x	x	x	x
Process non-compliant competitive quality fuel			x		x	x
Supervise perceived quality				x	x	x
Ensure benefits			x	x	x	x
Tracking possible improvements						x

Box 2.6. *Contributions of technical functions to principal functions*

This box shows that:

– the possible improvements are very different in nature and they can occur at any stage in the innovation process. Many technical functions can lead to improvements;

– after each process of improvement, it is necessary to carry out a capitalization of feedback in order to start any new innovation processes on new platform;

– the evaluation of the performance of the competitive quality process can take place at various stages in the innovation process and can take on various forms.

COMMENT 2.3.– The technical functions of a competitive quality process which qualify, validate, supervise, process non-compliance or ensure, are all processes which consist of the following stages:

– obtain factual information;

– ascertain the gap between the objective and the current situation;

– decide on how this gap can be closed;

– carry out this correction;

– record the process in order to capitalize on knowledge and expertise.

2.5. Conclusion

When we want to use quality as a tool to maximize the chances of an innovation process to succeed, it is necessary to fully understand what this tool is. This is the objective of this chapter, by showing a representative model.

In producing this model, we did not hesitate to offer precise definitions of the main concepts which generally characterize a competitive quality process within a value creation process.

These definitions may seem discouraging for a reader looking for principles to apply quickly in order to guide his processes. Experience shows that these definitions help in fully mastering the tool.

This global model of a competitive quality process integrated into an innovation process may seem maze-like because it has been deliberately placed within a TQM-type process, considering all relations that an innovation process undertakes with its various stakeholders.

In practice, the initial creation of competitive quality strategies for each class of stakeholder can help eliminate the need to oversee exchange relationships with some stakeholders in the competitive quality process. Common sense, habits and the importance attributed by some stakeholders to exchanges with the innovation project make the natural or implicit quality processes satisfactorily efficient.

The subsequent chapters will examine how to use this tool.

3

Competitive Quality Tactics

3.1. Introduction

In the following chapters, we will describe how a competitive quality process (CQP) can be used and how it functions in an innovation process to provide the best possible contribution to its success.

A CQP does not always function in the same way in the exchanges which take place with various stakeholders in the innovation process.

Firstly, it is necessary to distinguish between the exchanges with the future beneficiaries of the product or service which will incorporate the innovation, and the other stakeholders benefiting from traditional products or services, which can be repeated in every innovation process, and where only certain specifications need to be adjusted. However, for each stakeholder it is necessary to choose why to produce stakeholders' quality judgments on the products or services provided by the project for them to contribute to the project as the project manager wishes, despite the pressure of internal or external competition.

This is why it is necessary to choose specific competitive quality tactics (CQTs) for each class of stakeholder.

It is occasionally observed that competition is greater in exchanges with a class of stakeholders who are not direct beneficiaries of the products or services which will incorporate the innovation. This is the case when the innovation is intrinsically very attractive because it responds to an essential need which is not currently satisfied, and investors are difficult to attract because they are occupied with other projects, or they perceive the innovation as carrying high risk. In this case, the CQP must focus more on exchanges with investors than on those with future users of the products or services which will incorporate the innovation.

The following chapters are designed to provide a guide to how to use a CQP in exchanges with each class of stakeholder to ensure the success of the innovation. We have decided to structure them by successively addressing the various technical functions of CQPs named in the previous chapter (Box 2.6).

Each chapter successively addresses the objectives of the technical function, the function of the actions which produce it in the schedule of the innovation process, its specific actions, records of its performance and its managerial organization.

We will start with the design of the CQTs.

3.2. The objective of a competitive quality tactics

All exchanges in an innovation process with its stakeholders are based on socioeconomic interplay. The company gambles by offering certain products or services to its stakeholders hoping to obtain in return pre-defined contributions for the success of the innovation.

CQTs are a type of gamble often called: "quality speculation". The production of competitive quality fuel in the innovation process is a "stake" in the game. The uncertain events in the game are the stakeholders' competitive quality judgments during and after their exchanges with the company as the innovation process progresses. The winnings are the contributions of the stakeholders to the success of the innovation.

In-game tactics regard the choice of how to invest and play in order to obtain predetermined profits.

These tactics generally describe the choice of characteristics expected from stakeholders, and the products or services which the project must offer them in order to maximize the chances of obtaining resources in return. At this stage, they are very over-arching descriptions which will be specified by other technical functions of the CQP.

The purpose of CQTs can be defined in the following way:

> A CQT is intended to describe the main actions for producing competitive quality fuel of products or services offered to a stakeholder, and the benefits to be obtained in return during exchanges with the stakeholder, while encouraging him to commit to any other future innovation projects.

Box 3.1. *The purpose of competitive quality tactics*

This can be defined as a "win-win" operation.

COMMENT 3.1.– An entire book has been devoted to this single concept [MAI 13].

The project manager must design CQT in the exchanges which it plans to undertake with each class, or sub-class of stakeholder when a class is segmented. This is why it must firstly have an accurate map of the driver of socioeconomic exchanges in the innovation process.

We presented the main actions of CQTs in the previous chapter. We shall recall them here.

CQT associated with exchanges with a stakeholder:

– describes the main characteristics of the competitive quality judgments on the product or service offered to stakeholders, to encourage them to provide the resources expected by the project manager, in the most profitable way possible;

– defines the main characteristics of the compensatory resources which allow the project to reach its objectives, ensuring sustainable contributions from the stakeholder;

– chooses the most profitable processes and resources to implement within the project in order to prompt competitive quality judgments and benefit from the compensatory resources.

Today, an innovation project is potentially in competition with each of the following classes of future stakeholders:

– a worker, earmarked to be a member of the project team, is often required to work on other projects simultaneously. Why would he put more energy into the innovation process than the other projects?

– a company, specialized in "capital risk", has the choice between several projects. Why would it invest in the innovation project?

– public authorities select innovation projects to award them with funding;

– at the moment of choosing the investments which will allow it to develop, the company often needs to choose between several innovation projects;

– the future providers of the project, or the business unit (BU) which will produce the products or services to incorporate the innovation, are in turn obliged to invest in order to provide the resources required. This is a risk which they will only take if the project appears to be more profitable than other investments;

– why would future clients for the products or services which will integrate the innovation be more attracted by these than by the competitors' offer?

The project manager must therefore predict the competitive quality judgments of the products or services which will be offered in order to fend off the competition, for each potential class of stakeholder.

It is in accordance with the attractiveness of these judgments that the characteristics of the products or services to be requested from the stakeholder in return can be established, without provoking a refusal from the part of the stakeholder.

Finally, it is in accordance with these predictions that actions must be envisaged and the most profitable resources to be implemented in order to encourage competitive quality judgments and obtain these resources.

To carry this out, it is essential to:

– have a clear view of the evolutions in the driver of socioeconomic exchanges linked to the innovation during the process (Box 1.11, Chapter 1);

– ascertain the main characteristics of the support expected from stakeholders in the innovation process in order to achieve success;

– have good knowledge of the possible uses of quality in an innovation process to obtain the support expected from the stakeholders (Chapter 2).

The design of such tactics almost always exists in an empirical and often implicit way. The provision of a "business plan" to future investors is subject to often implicit CQTs. The project manager considers how he will present this document to effect the competitive quality perceptions of investors in order to obtain their financial contribution to the desired level, what resources he will be able to use to carry out this presentation and so on.

The design of CQT is often considered, by project managers, as a burdensome constraint. They struggle to make predictions for the future given the scale of the uncertainties which are attached to their innovation project. It is often perceived as a reduction of the freedom necessary for the creation of value in the innovation process. They frequently feel obliged to consider questions at the wrong moment. However, this process is now absolutely vital. It must be taken into account by the project manager, as well as the quality service which accompanies it, throughout the innovation process. CQTs can undergo changes at any time. These changes must be accompanied by *configuration management*.

When an evolution occurs, configuration management of such an evolution consists of:

– providing understandable guidelines for new objectives to be met;

– describing, and interpreting, the gaps between these objectives, the results already obtained, and the previous objectives;

– setting out the rationale for these new objectives.

3.3. Planning competitive quality tactics in the innovation process

Generally speaking, as soon as the project manager realizes that services from a stakeholder are necessary for the correct development of an innovation process, he must design relatively precise tactics to make the best use of quality in these future exchanges. To do so, he must have a clear understanding of the driver of socioeconomic exchanges in the innovation process (Box 1.11, Chapter 1).

CQT must be gradually constructed before, and during, the time when contacts are being made with a stakeholder. This will allow it to establish what it expects from the stakeholder, as well as its attitudes and behavior, and its ability to produce the support that the company requires from the stakeholder.

At any time, CQT can be modified in accordance with feedback received, but also due to any evolutions in the demand for products or services released by the project manager, or expectations from the stakeholder.

Consequently, CQT integrated into an innovation process must be extremely flexible and remain over-arching.

The moment when CQT can be designed can be established in accordance with the following model, which is a direct continuation of Box 1.1 given in Chapter 1.

Lines of action	Stages of the innovation process
Creative research	– Bringing about the emergence of an idea which may provide a response to a need of society; – Creating experimental drafts which provide a more accurate representation of the idea; – Progressively designing drafts to accurately represent the configuration of the innovation; – Progressively designing drafts which give a representation of future products and services to incorporate the innovation, by describing, in particular, the future "interfaces" of the innovation with other components of these products and services.
Socioeconomic impact studies	– Briefly characterizing products or services which could benefit from the innovation; – Identifying possible clients for these future products or services; – Characterizing future uses for products or services which would incorporate the innovation; – Conducting prospective studies on possible socioeconomic uses for the innovation, and its impacts on the company and its position; – Identifying stakeholders who will contribute to the development and production of the future products and services which will incorporate the innovation; – Conducting tests on the drafts of future products or services, at various steps in their production, on future clients and on all future stakeholders in the production system for the products and services.

Feasibility studies	– Conducting initial feasibility tests on the future products or services; – Identifying future stakeholders in the innovation project; – Conducting studies on the nature and feasibility of exchanges with future stakeholders in the project necessary for it to be fully realized; – Conducting performance tests on future products or services; – Conducting studies on the ability of the company to design production and distribution systems for the products and services which will incorporate the innovation by taking into account the configurations of the various definitive drafts of the innovation.
Organization	– Building a processor responsible for carrying out the innovation process and an external communication network; – Creating a rationale to aid decision-making, examining the probable socioeconomic impacts of implementing the innovation; – Designing prototypes of the production and distribution systems for the products and services; – Creating a rationale to aid decision-making which details the feasibility of implementing the innovation; – Examining the steps to take to ensure the intellectual protection of the innovation.
Decisions	– Deciding to foster new ideas to help the progress of the company; – Deciding to initiate prospective studies on how to use the idea to benefit the company; – Deciding to guarantee the ability of the company to mobilize the resources necessary, internally and externally, in order to fully realize its innovation process;

	– Deciding contractual issues with these stakeholders, and building a processor to carry out the innovation process; – Deciding to progressively design drafts which give a more accurate representation of the idea; – Deciding to carry out prospective studies on the socioeconomic impacts of the future products and services which will incorporate the innovation; – Deciding to guarantee the ability of the company to adopt a production system for the future products and services; – Deciding to validate the innovation.

Box 3.2. *Stages and lines of action in an innovation process*

We start with the CQT which deals with relations with the end users of the future products or services to incorporate the innovation. The first stages of the CQT's design take place during the following stages of creative research:

– bringing about the emergence of an idea which may provide a response to a need of society;

– creating experimental drafts which provide a more accurate representation of the idea.

As the design of the CQT continues, it is extended to relations with other future stakeholders who will use the products or services incorporating the innovation. For instance, these stakeholders may be the distributors of the products or services, or economic representatives who will include these products or services in a wider offer to consumers.

The design of these CQT takes place during the following stages of creative research:

– progressively designing drafts to accurately represent the configuration of the innovation;

– progressively designing drafts which give a representation of future products and services to incorporate the innovation, by describing, in particular, the future "interfaces" of the innovation with other components of these products and services.

The design of these CQT is continued in the following stages:

– identifying possible clients for these future products or services;

– conducting prospective studies on possible socioeconomic uses for the innovation, and its impacts on the company and its position;

– characterizing future uses for products or services which would incorporate the innovation.

During these various stages, it is necessary to try to identify the future commercial target of the innovation. It is common at this stage to see what is often called "the splits" because it combines the two extremes in the process: the idea and the future clients of the "business unit" which may possibly produce the products or services incorporating the innovation.

This comes to an end with the stage focused on:

– conducting performance tests on future products or services.

Subsequently, the project manager must accurately define the CQT that he wishes to adopt with "investors" who will participate in funding the project. This includes the creation of a "business plan".

From this stage, the CQT no longer deals with the products or services which will incorporate the innovation, but with other more traditional products or services, specific to each class of stakeholders, that the projects must provide in order for them to agree to collaborate with the innovation process.

The following CQT addresses the class of stakeholders made up of project team members. Such a CQT depends on the internal competition which exists between different projects which are being developed at the same time. It is during these processes that the project manager may decide to surround himself with workers with certain skills in order to help the project evolve. This third design stage for CQT includes other stakeholders who contribute to the innovation process (stakeholders in the strategic area of activity).

The stages of the innovation process which integrate the design of these tactics are as follows:

– conducting prospective studies on possible socioeconomic uses for the innovation, and its impacts on the company and its position;

– conducting initial feasibility tests on the future products or services;

– identifying stakeholders who will contribute to the development and production of the future products and services;

– conducting studies on the nature and feasibility of exchanges with future stakeholders in the project necessary for it to be fully realized;

– building a processor responsible for carrying out the innovation process and an external communication network.

At this stage, CQT for desired exchanges with future designers and producers of the products and services which will integrate the innovation should not yet be started.

This is drawn up during the following stages of the innovation process:

– conducting prospective studies on possible socioeconomic uses for the innovation, and its impacts on the company and its position;

– identifying stakeholders who will contribute to the development and production of the future products and services;

– conducting performance tests on future products or services;

– designing prototypes of the production and distribution systems for the products and services;

– conducting studies on the ability of the company to design production and distribution systems for the products and services which will incorporate the innovation by taking into account the configurations of the various definitive drafts of the innovation;

– conducting tests on the drafts of future products or services, at various steps in their production, on future clients and on all future stakeholders in the production system for the products and services.

The project manager must finally produce CQT in relations with the company directors to convince them to operationally commit to the innovation by entrusting a BU to the design, production and distribution of the future products or services which will incorporate it. It is designed during the following stages of the innovation process:

– creating a rationale to aid decision-making, which examines the probable socioeconomic impacts of implementing the innovation;

– creating a rationale to aid decision-making which details the feasibility of implementing the innovation.

Throughout the stage of gradual draft design specifying the innovation configuration, the project manager must retain the possibility of revising the CQTs in order to adjust them to the new resource needs of the project, and to take into account the evolution in the exchanges with various classes of stakeholder in the processor carrying out the innovation process.

This flexibility in the configuration of CQT is essential. Each configuration must be based only on information known by the project manager, which gives credibility to the predictions on the future of the innovation process.

3.4. Methods for designing competitive quality tactics

As we have seen above, CQT is refined as the innovation process progresses.

Designing CQT is a process. As for any process, it is necessary to start by specifying input data or initial conditions which make up the basis of constraints and opportunities which will help to create the CQT and facilitate the feasibility study.

The main areas of input data in the process of designing CQT can be described as follows:

– description of macro-characteristics of demands or expectations of the stakeholder regarding the products or services which have already been offered, or that must be satisfied in order to benefit from its services;

– identification of CQT of the main competitors which may disrupt exchanges with the stakeholder.

Subsequently, the project team attempts to establish a hierarchy of the types of quality perceptions that must be prompted by highlighting differences in comparison with the competition, so that the stakeholder makes a competitive quality judgment on the products or services he will receive from the project. This process is a kind of marketing macro-study. The descriptions of perceptions to be prompted should be limited to very general statements.

The process consists of:

– identifying the expectations not being satisfied by the competition, to which the stakeholder attaches a great deal of importance, and that the project is able to satisfy better;

– discovering the stakeholder's behavioral mechanisms which are at the basis of these quality perceptions in terms of satisfying expectations, and which prompt a competitive quality judgment (these behavioral mechanisms are later called the "oxidizer");

– describing and establishing a hierarchy of quality perceptions in accordance with the importance of their impacts on its competitive quality judgment.

The search for this information must be carried out as naturally as possible. It generally relies on the company's staff members who are most often in contact with the stakeholder. This process is the same regardless of the class of stakeholders in question.

Often, the stakeholder standardizes its expectations in a set of specifications.

At the same time as this process, the project manager must describe the main characteristics of the services that he expects to receive from the stakeholder in exchange for the products or services which it is offering (expected benefits).

These services must contribute to the success of the innovation process in the short or medium term.

The characteristics of these services must be designed to take into account:

– the ability of the stakeholder to produce them, and the ways the project will be able to use them;

– the perceived interest from the stakeholder to provide the work requested in order to obtain the products or services offered by the project manager.

The estimates of these benefits are predictions. They can include a time-based element, and, when benefits are quantifiable, they can be presented in the form of a confidence interval [MAI 13].

The following stage in the design of the CQT regards the very general characterization of the quality fuel of the products or services which are offered to the stakeholder in return for its services.

For instance, if the stakeholder is a financial organization, information contained within provisional balance sheets is part of the quality fuel of the products or services offered to this organization. This information must satisfy the specifications formulated by this organization.

Although it is sometimes difficult to update the behavior of the stakeholder who transforms the quality fuel from a product or service into a competitive quality perception, it is easier to identify the nature of the information which makes up the quality fuel of the product or service which are offered. The stakeholder generally communicates willingly on this subject, and this often constitutes part of the specifications sent to the project manager.

The final stage of designing CQT consists of carrying out a global feasibility study on the release of information which would constitute the future quality fuel of the products or services.

To do this, the project manager, assisted by his team members, identifies the primary information which will be used to produce the quality fuel and establishes the main characteristics of the future configurations of the quality fuel emitters. These emitters must be integrated into the processor which will carry out the process.

COMMENT 3.2.– At each stage of the process to design CQT, it is possible to go back and modify the output data from the previous stage. This future prediction carries so much uncertainty that it is essential to allow for this amount of freedom.

CQT can also, at any time, be modified in accordance with the stakeholder's observed behavior.

For more operational details regarding the design of CQT, we recommend our book entitled *Competitive Quality Strategies* [MAI 13].

3.5. The performance of competitive quality tactics

The performance of CQTs is gauged by the quality judgment of the project manager on the services provided by the stakeholder. This judgment is based on the quality perceptions of the project manager and his team as defined in Chapter 2.

It is very important to take stock of this quality judgment during each project review stage, in order to establish a diagnosis of the quality tactics, if necessary, to carry out corrective action as quickly as possible.

3.6. Managing the design of competitive quality tactics

The design of CQT is a process carried out under the responsibility of the project manager. He will lead the process.

The managerial organization of the process can be described with the help of the following model:

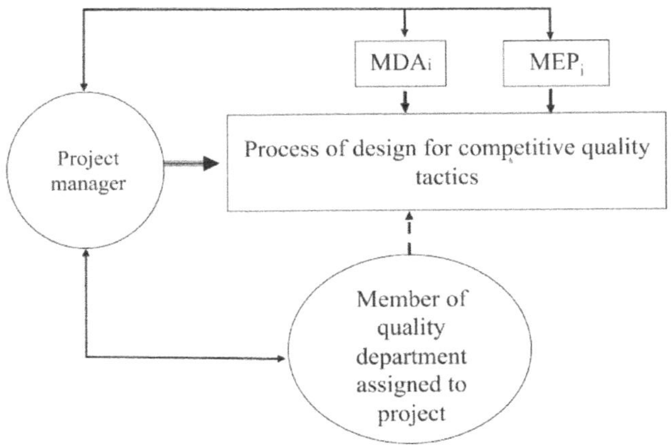

Figure 3.1. *Managerial organization of the processor for the design of a CQT*

COMMENT ON FIGURE 3.1.–

MDAi denotes the members of the management within the company responsible for relations with the class of stakeholders affected by the CQT (human resources management, financial service management, marketing management, purchasing management, administrative management, legal department, etc.).

MEPi denotes the members of project team number (i) who contribute to the added value of the innovation process by supplying a particular type of skills related to an area of expertise.

Relations between project manager and other members of the processor are *managerial relations by delegation*. The various members who participate alongside the project manager are seconded to the project. They belong to other "managerial areas" and are made available for the sake of the project. The project manager gives orders to these members and provides the resources for them to carry such orders out. These relations must take place in a strong spirit of partnership, whereby the project manager and these members must work together in accordance with the order statement, the definition of the resources to be assembled, and these members can request assistance from the project manager in carrying out such tasks.

The initial meetings are generally focused on ensuring that the processor is able to draw up a CQT. The project manager checks that the team members share the opportunity and feasibility of such a CQT design process. He also validates the process set-up, distributing roles among the various members.

The project manager then decides to go ahead with the CQT design process.

During the entire process, the project manager oversees the ability of the working group to jointly produce the expected results, polices any deviations, checks the relevance of information released and ensures that decisions are shared.

He requests that the quality department propose a system for assessing the risks of not reaching objectives required for the design of CQT and for carrying out the necessary alerts.

He makes decisions on corrective action depending on the problems observed.

Throughout the process, the project manager must show the interest he attaches to the CQT design through his behavior and attitude. He must show that the information released will produce real added value to give rise to competitiveness factors. He must also show that he is ready to overcome the challenges which have emerged. It is essential that the motivation of the processor's human resources is maintained at a high level throughout the process.

At the end of the process, the project manager is to carry out a report on the initial relevance of the CQT, the level of consensus in the choices made and the ability of team members to then work together in carrying out the CQT.

3.7. Conclusion

The design of various CQT to be implemented in an innovation process may seem tedious. It requires the project team to devote its time to the relations which it must forge with numerous company stakeholders involved in the process. However, the socioeconomic success of the innovation depends on it. It is not a question of it being torn apart by the various stakeholders, but, on the contrary, of capitalizing on the attractive potential of the company and the project to encourage them to supply the process with the resources to ensure its success. The potential perceived quality of the company and project will produce this attraction.

The members of the project team cannot carry out creative research and discuss the relations that will be forged with stakeholders at the same time. It is necessary to alternate between these two logical steps.

The design of CQT is the start of a process of staging how the exchanges between the company and some of its stakeholders shall take place. This staging is much more complex than in a theater production because it involves a greater number of influencing factors, and the members retain a greater freedom for improvisation.

The design of CQT is at the intersection between the traditional definitions of compliance quality objectives and marketing studies generalized for all stakeholders.

The speed and accuracy in how the design of CQT is carried out can vary depending on the level of pressure from competitors on stakeholders and the ability of capitalizing on feedback regarding products or services offered to stakeholders.

4

The Marketing Quality of an Innovation

4.1. Introduction

Defining competitive quality tactic (CQT) means establishing a model for co-producing quality with a class of stakeholders in an innovation process.

The question which arises is:

> *How can this co-production be carried out while ensuring that the CQT provisions are complied with?*

The objective of this chapter is to illustrate the main events which ensure that this co-production complies with these provisions.

It is a matter of visualizing the key characteristics of these key events which must take place during exchanges between the innovation project and the members of a class of stakeholders to ensure compliance with the CQT.

For instance, this includes anticipating the most important events which may occur in negotiations between the project manager and project investors, so that they perceive certain aspects of competitive quality within the innovation which motivates them to provide sufficient financial support for the needs of the project.

Events also need to be anticipated in future exchanges between the project manager and the business unit (BU) responsible for integrating the innovation into a new product so that they perceive certain aspects of competitive quality within the innovation motivating them to carry out this integration.

These anticipations are possible because there are implicit or explicit socioeconomic "conventions" in relations with these various parties.

This description of events does not make assumptions on how the competitive quality fuel (CQF) intended for stakeholders will be produced.

Instead, it will consist of a collection of opportunities or constraints from the set of specifications imposed on the design process for this quality fuel and its release.

An innovation is a "one-shot" process. Usually, a company has little usable past experience apart from information regarding the main methodological principles or technological or organizational similarities, to guide the design and operation of the processor in carrying out the innovation.

Answering this question of "how?" in this design stage is therefore essential. It may appear even more complex when a number of different stakeholders are involved. In reality, this stage takes place over an extended time period and is distributed among various competences, apart from in the cases of particularly small project teams.

Here, we present a number of methodological principles applied to the implementation of a single set of CQTs. They can be generalized across any combinations of CQT relating to a mixture of stakeholders.

The only extra complications arising from this generalization come from the fact that a set of CQTs which applies to a class of stakeholders can reduce the freedom in designing other CQT.

4.2. Marketing quality in an innovation process

"Marketing quality in an innovation process" refers to the creation of scenarios in some processes to guide the implementation of various CQT provisions.

These processes are:

– the release of the CQF of the products or services offered to the stakeholders in the innovation process;

– the stakeholders' expressions of competitive quality judgments;

– the supply of compensatory resources from the stakeholders to the project, or to the future BU;

– the contributions of stakeholders to the project image and the company's reputation.

These are cross-cutting processes for the "production/use" of quality within an innovation process which are involved in all exchange relationships with any class of stakeholders.

In this case, the term scenario is:

> *A description of a pre-defined and organized flow diagram of an action.*

To design or envisage the scenario of a process is to describe how certain key events in the process must take place and trigger others. The choice of these particular events is based on how they guide those carrying out the process and on the importance of their contributions to the process' end result.

A scenario can then help to maintain the process under control. In particular, the description of contributions of these events to the end result of the process limits the risks of not meeting its objectives. Process' marketing quality is often confused with their "cartography". The objective is the same: design guidelines for the process by anticipating its development. However, in this case, we are focusing on only four types of process within the wider innovation process, located at the heart of the production and use of quality in compliance with the CQT provisions, whereby certain key events are described but not the organization of the bodies which produce them. Some specifications imposed upon these bodies may possibly be described because they release information which makes these events explicit. These key events are often attitudes or behavior of the parties contributing to the development of the process.

For instance, the description of key events to produce the specifications of a part imposed by the set of specifications negotiated with the client is this type of scenario. Envisaging how this production will take place means envisaging the events which appear the most essential to the production, their planning and a general representation of the architecture of the processor which will carry it out. These events express a "how to", shared among the future members of the production process. The ways that these events are described must be chosen by process designers and they must be formulated in a language which allows future team members to understand and implement them. The designers can use a methodology which facilitates the creation of these scenarios.

In the catering sector, presenting a menu to a client is a visible process: for the client in question, the other clients and those serving in the restaurant. Predicting the scenario of events which will have an influence on the quality perceptions of these stakeholders requires certain aspects to be described: the appearance of the menu, the way in which the waiter must present it and engage with the client, and the way in which he must lead the exchange of information with the client so that he is satisfied with the service, while maintaining the satisfaction of the other staff members in the restaurant.

The creation of this type of scenario is part of traditional work in the marketing department when stakeholders are clients. Here, we are simply describing this work in a different way, and generalizing it to all stakeholders.

These principles must be applied to the four types of specific processes above which ensure the best use of quality in any innovation process.

These scenarios have a relative amount of precision because they are not guidelines. Just as in a theatre production, they leave room for improvisation.

Some lines of action described in these scenarios are compulsory, others are rules which take the form of: "if … then …". They sometimes leave a relatively significant amount of room for improvisation, in order to produce some process events in unpredictable situations or to facilitate creativity in the innovation process.

The CQT influences the marketing quality in these processes in the following way:

– it outlines the target stakeholders;

– it gives indications of the needs or expectations of stakeholders who should be satisfied as a priority;

– it focuses the design of scenarios on specific areas of stakeholders' quality perceptions and on their behavior and attitudes which are key components of their quality fuel;

– it describes the benefits to be obtained from the compensatory resources supplied by the stakeholders;

– it focuses attention on certain specific functional characteristics of the products or services offered to stakeholders, to direct the design of those scenarios relating to how the CQF is released so as to prompt competitive quality perceptions;

– it provides general rules to be implemented in exchanges with certain providers of processors producing the products or services offered to stakeholders.

By envisaging how these relations take place between the project and the stakeholder, how the stakeholder reacts on discovery of the project or service, and how it obtains and uses it, we are able to design scenarios for this process.

The project manager may plan a number of scenarios.

This predictive description of how the exchanges will take place between the project and the stakeholders affects the way other technical functions of the competitive quality process (CQP) integrated into the innovation process are carried out.

If a scenario does not occur as predicted by the members of the project team who observe deviations, the project manager must be alerted soon enough in order to make the necessary modifications.

To design these scenarios, the project manager must take into account a large number of parameters and predict the value which will be assigned to them by stakeholders during exchanges. This prediction into an uncertain future is made with risks which are difficult to estimate since history never repeats itself exactly. Nevertheless, this risk is part of the constraints of any innovation process. It is limited by the reactiveness of the project team when a deviation is spotted.

The main reasons for failure in an innovation process can be attributed to its marketing quality.

This function of the CQP plays a key role in the success of an innovation.

4.3. Planning marketing quality within the innovation process

In the same way as in the previous chapter, our approach is based on modeling the various stages of the innovation process, and reviewing the four types of processes discussed in the previous section:

– CQF releases for the products or services offered to the stakeholders in the innovation process;

– forming stakeholders' competitive quality judgments;

– provision of compensatory resources from stakeholders for the project, or the future BU;

– stakeholders' contributions to the project's image or the company's reputation.

4.3.1. *Releases of competitive quality fuel*

The following box shows the stages of the innovation process in which the scenarios for CQF emissions must be created according to the class of stakeholders they relate to.

Category of stakeholders	Stages in the innovation process
Future users of products or services incorporating the innovation	– Creating experimental drafts which provide a more accurate representation of the idea; – Conducting prospective studies on possible socioeconomic uses for the innovation, and its impacts on the company and its position; – Progressively designing drafts to accurately represent the configuration of the innovation; – Progressively designing drafts which give a representation of future products and services to incorporate the innovation, by describing, in particular, the future "interfaces" of the innovation with other components of these products and services; – Briefly characterizing products or services which could benefit from the innovation; – Characterizing future uses for products or services which would incorporate the innovation; – Conducting tests on the drafts of future products or services, at various steps in their production, on future clients and on all future stakeholders in the production system for the products and services; – Conducting performance tests on future products or services; – Designing prototypes of the production and distribution systems for the products and services.
Potential investors in the innovation and other stakeholders contributing to the project	– Conducting initial feasibility tests on the future products or services; – Conducting prospective studies on possible socioeconomic uses for the innovation, and its impacts on the company and its position; – Identifying future stakeholders in the innovation project;

	– Conducting studies on the nature and feasibility of exchanges with future stakeholders in the project necessary for it to be fully realized; – Building a processor responsible for carrying out the innovation process and an external communication network.
Future contributors to the "BU" which will produce and distribute the products or services incorporating the innovation	– Conducting prospective studies on possible socioeconomic uses for the innovation, and its impacts on the company and its position; – Identifying stakeholders who will contribute to the development and production of the future products and services which will incorporate the innovation; – Conducting performance tests on future products or services; Designing prototypes of the production and distribution systems for the products and services; – Conducting studies on the ability of the company to design production and distribution systems for the products and services which will incorporate the innovation by taking into account the configurations of the various definitive drafts of the innovation; – Conducting tests on the drafts of future products or services, at various steps in their production, on future clients and on all future stakeholders in the production system for the products and services.

Box 4.1. *Marketing quality in the processes to release the competitive quality fuel of an innovation*

Marketing quality in the process of releasing CQF is involved in planning the largest number of stages in the innovation process. The creation of these scenarios is an essential part of the innovation process.

4.3.2. *The emergence of stakeholders' competitive quality perceptions*

The previous process is reproduced. The stages of the innovation process where the scenarios are created in order to give rise to competitive quality perceptions from various classes of stakeholder are shown in Box 4.2.

Category of stakeholders	Stages in the innovation process
Future users of products or services incorporating the innovation	– Conducting prospective studies on possible socioeconomic uses for the innovation, and its impacts on the company and its position; – Characterizing future uses for products or services which would incorporate the innovation.
Potential investors in the innovation and other stakeholders contributing to the project	– Conducting prospective studies on possible socioeconomic uses for the innovation, and its impacts on the company and its position; – Conducting studies on the nature and feasibility of exchanges with future stakeholders in the project necessary for it to be fully realized.
Future contributors to the "BU" which will produce and distribute the products or services incorporating the innovation	– Conducting prospective studies on possible socioeconomic uses for the innovation, and its impacts on the company and its position; – Conducting tests on the drafts of future products or services, at various steps in their production, on future clients and on all future stakeholders in the production system for the products and services.

Box 4.2. *Marketing quality of the process giving rise to stakeholders' competitive quality perceptions*

During these stages of the innovation process, intended to better understand the consumption attitudes and behavior of stakeholders regarding the various products or services created by the project, scenarios can be defined which deal with the emergence of stakeholders' competitive quality perceptions.

4.3.3. *Compensatory resources and contributions to the company's image and reputation*

In the same way, the stages of the innovation process regarding scenarios which deal with the provision of compensatory resources from the various classes of stakeholder and their contributions to the company's image and reputation are shown in Box 4.3.

The scenarios regarding this aspect of marketing quality are produced at the same time as the stages of the innovation process which deal with the marketing quality of the emergence of competitive quality perceptions.

Category of stakeholders	Stages of the innovation process
Future users of products or services incorporating the innovation	– Conducting prospective studies on possible socioeconomic uses for the innovation, and its impacts on the company and its position; – Conducting performance tests on future products or services.
Potential investors in the innovation and other stakeholders contributing to the project	– Conducting prospective studies on possible socioeconomic uses for the innovation, and its impacts on the company and its position; – Conducting studies on the nature and feasibility of exchanges with future stakeholders in the project necessary for it to be fully realized.

Future contributors to the "BU" which will produce and distribute the products or services incorporating the innovation	– Conducting prospective studies on possible socioeconomic uses for the innovation, and its impacts on the company and its position; – Conducting tests on the drafts of future products or services, at various steps in their production, on future clients and on all future stakeholders in the production system for the products and services.

Box 4.3. *Stages in an innovation process where the marketing quality of the processes for the provision of stakeholder's compensatory resources is defined*

4.4. Defining marketing quality

The previous boxes show that the creation of various scenarios which constitute the marketing quality of an innovation process always has two stages. The first is intended to envisage a scenario; the second aims to check its credibility with a test or a very limited simulation. Three main reasons limit these tests: their cost, the deadline for putting an innovation on the market and the necessary confidentiality to prevent the competition being supplied with information.

Envisaging this type of scenario is exactly the same process as creating a scene in a film.

We have seen that a process is carried out by an "events driver" called a "processor".

Marketing quality does not attempt to describe the processors of the four processes for the "production/use" of quality within an innovation process. It is limited to modeling the key events of these four processes.

For each of these, the first step is to describe how the key events that the project must implement during the process are represented. This a description of the "what to do" step.

This representation is created by members of an innovation process who must:

– take into account feedback from previous experience;

– take into account knowledge regarding stakeholders' consumption methods of a similar product or service to the one the project is planning;

– integrate any demands made in the specifications sent by stakeholders;

– integrate demands featured in standardization, safety and environmental protection regulation;

– make use of known socioeconomic conventions specific to the company's sector, which take into account stakeholders' culture and the company members participating in the process;

– comply with the provisions in the CQT.

Capitalizing on feedback facilitates this process. This is why it is advised that a project promoter should seek assistance from an incubator. Today, there is support available for other types of innovation project in "competitiveness clusters".

As the previous boxes highlight, these processes are repetitive. Some of these become more precise during the innovation process.

The creation of these scenarios allows certain tasks carried out by the innovation project team to be planned.

Subsequently, the second part of the scenarios deals with the key events which will be carried out by the project's stakeholders. Indeed, the four quality "production/use" processes are co-produced by the project and the stakeholders. The scenarios anticipate this second type of events, although they cannot be completely controlled by the innovation process. These events must be considered as constraints or opportunities to carry out those which are produced by the project.

At this stage, the risk is such that it is necessary to carry out feasibility studies of these scenarios. As shown in the boxes above, this means carrying out very limited tests. This method of filtering the criticality of the risks depends on the prior identification of these risks and the accumulation of information during the innovation process which causes their initial criticality to evolve. Some project reviews must be used to validate these feasibility studies, or to decide to modify the scenarios.

The main stages in creating the various "marketing quality" scenarios of a CQP can be modeled as follows.

4.4.1. *Marketing quality in releasing competitive quality fuel*

The marketing quality in releasing CQF with regard to a given class of stakeholders consists of:

– identifying the events of the products or services, produced by the project team or co-produced with stakeholders, which release information prompting competitive quality perceptions, while complying with the provisions of the CQT;

– describing a pre-defined and organized flow diagram for each of these events;

– presenting this diagram to certain stakeholders to gauge their reactions;

– interpreting their reactions in terms of opportunities or risks with regard to the effects of their competitive quality perceptions, the provision of compensatory resources to the project or future BU, and their contribution to the company image or reputation;

– modifying the initial diagram to comply with provisions in the CQT.

4.4.2. *Marketing quality in the emergence of competitive quality judgments*

The marketing quality of the events encouraging the stakeholder to make a competitive quality judgment on the products or services offered by the innovation project consists of:

– identifying the events which may be observed that show how the information of the CQF of the project's products or services offered to stakeholders will have an impact on their competitive quality perceptions;

– describing a probable flow diagram for these events;

– testing this flow diagram with some stakeholders to observe the reactions to a simulation of a scenario releasing the CQF of the products or services;

– characterizing stakeholders' competitive quality judgments which arise from these tests;

– comparing these judgments with provisions from the CQT;

– modifying the initial flow diagram, where necessary, as well as the related CQF production scenario in order to comply with the provision of the CQT.

The identification of the observable events of the impacts of CQF on competitive quality perceptions can be guided by the following methodological principles:

The goal is first to identify how the CQF information is obtained by stakeholders. Their input is filtered; therefore, it is appropriate to characterize the key events in the transformation of this information in terms of inputs.

Subsequently, it is necessary to identify the events which take place during the interpretation of the transformed information.

To carry out such interpretation, the stakeholder needs to:

– move this transformed information closer to its own experience;

– compare this information to that produced by the CQF of rival products or services;

– move this information closer to its expectations.

These comparisons always represent the estimated difference between the observed reality and the predicted one.

Finally, the stakeholder assigns an intrinsic value to each one of these differences.

The estimated differences help to give rise to competitive quality perceptions. The values attributed to these differences lead to those which are attributed to the various perceptions.

It is these perceptions and their values which contribute to each stakeholder's competitive quality judgment.

"Stakeholder's oxidizer quality", the system: receiving, interpretation, and recovery, which filters the CQC information of company products or services, in order to produce its quality perceptions.

The “quality oxidizer” is different for each stakeholder. However, a section of these stakeholders is specifically based on characteristics which are common to their CQFs. When the stakeholders are clients, the marketing department is responsible for finding out the general operational mechanisms of clients’ CQFs.

4.4.3. *Marketing quality in the production of compensatory resources*

Stakeholders agree to make the desired contributions to the innovation process if they believe that their competitive quality judgment of the products or services which they will receive justifies such a contribution.

Drawing up a scenario of these contributions consists of:

– identifying observable events which lead to a judgment being made on the compliance with the initial specifications of the stakeholders’ contributions to the innovation process;

– describing a likely flow diagram of such events;

– testing this diagram with future users of these contributions within the innovation process;

– estimating the chances of whether the fulfillment of this scenario will lead to a successful innovation process;

– carrying out any necessary modification of this scenario and specifications of contributions requested from the stakeholders in order to comply with the provisions of the CQT and strengthen the chances of success of the innovation process.

This component of marketing quality is relatively easy to produce since the events in question involve issues which are traditionally dealt with by the project teams of innovation processes.

4.4.4. *Marketing quality in contributing to the company's image and reputation*

The stakeholders in an innovation process, even if they are restricted to certain levels of confidentiality, are often driven to make a more general judgment about the company's performance, with the information that they exchange during contacts with the company or its members. These judgments inevitably contribute to the image or reputation of the company, and more specifically reveals a company's ability to successfully carry out an innovation process.

Drawing up a scenario of these contributions consists of:

– identifying observable events which lead to a judgment being made on the compliance with the initial specifications of the stakeholders' contributions to the innovation process;

– describing a likely flow diagram of such events;

– testing this diagram with future users of these contributions within the innovation process;

– estimating the chances of whether the fulfillment of this scenario will lead to a successful innovation process;

– carrying out any necessary modification of this scenario and specifications of contributions requested from the stakeholders in order to comply with the provisions of the CQT, and strengthen the chances of success of the innovation process.

Stakeholders agree to make the desired contributions to the innovation process if they believe that their competitive quality judgment of the products or services which they will receive justifies such a contribution.

Feedback analysis plays a key role in the production of this component in the innovation's marketing quality.

4.5. Marketing quality performance

The first level of marketing quality performance in an innovation process affects the way in which the various scenarios are carried out during exchanges with stakeholders. They may experience unexpected influences from phenomena in the exchange environment, or be modified by unexpected behavior from company members.

Observing instances of non-compliance of marketing quality events to their scenarios constitutes an initial level of performance.

The second level of performance regards the effects of the events targeted by marketing quality on the real contributions of stakeholders to the success of the innovation process and the profitability of the energy expended in order to produce the products or services offered to stakeholders who agree to provide such contributions.

For instance, if the project's financial resources allowed it to be fully completed and if the result was obtained in compliance with the predicted scenarios for carrying out the following four processes:

– releasing the CQF of the products or services offered to the project's financial backers;

– competitive quality judgments of financial backers on these products or services;

– contribution of financial backers to the project's funding;

– contributions of financial backers to the company's reputation regarding its ability to offer good investment prospects.

Then it could be said that the competitive marketing quality regarding exchanges with financial backers was effective.

However, if financial resources are seen to be insufficient for completing the project and it is necessary to find new financial backers to conclude the project, either because the scenarios were complied with but were poorly designed, or because the scenarios could not unfold as expected, then marketing quality was ineffective.

For example, a project manager was able to design a scenario with a potential financial backer, believing that it was able to provide a significant portion of the project funding. However, in reality, this stakeholder did not have the necessary resources to make this financial contribution despite the high level of interest it expressed in the project. In this case, the scenarios of the first two processes were completed as planned, but the third process could not be fulfilled.

Furthermore, if the financial backer has the resources to make a sufficiently important contribution to the project, but it has not been persuaded of the likely profitability of its investment and has only committed partial support to the project, the scenario of the second process, regarding stakeholder's quality perceptions, has been poorly designed or has not been complied with.

In this example, the project manager may also observe that some of the scenarios, regarding events produced by the project team, lead to pointless expenditure. In this case, marketing quality is not optimal because some expensive events, which do not impact the success of the innovation, did not have to feature in the scenarios. Such observations often help to simplify the scenarios by taking advantage of feedback and making improvements on productivity while making the process more robust.

The critical analysis of the competitive marketing quality of an innovation process often helps to make significant improvements in productivity and make the contributions of various stakeholders easier.

These examples show that it is not necessary to implement a quantitative measure of marketing quality performance in order to appraise it and develop any necessary improvements.

As displayed above, marketing quality is a prediction on how the four specific types of process will be carried out as part of an innovation process. We can produce what is also often referred to the desired "event pathway" of these processes.

Therefore, during the experimental and real stages of these processes, it is possible to observe any deviations from these pathways and estimate, or observe, the consequences of these deviations on the success, or chances of success, of the innovation. This type of recognition during the project can help to adjust the marketing quality as the project progresses. They are generally decided upon during the project reviews.

4.6. Managing the production of marketing quality

The production of marketing quality in an innovation process, with regard to a specific class of stakeholders, is the responsibility of the project manager, who is in charge of leading this process.

This process, which consists of designing scenarios of other processes, uses specific expertise which is enhanced with the help of feedback. When the innovation process is repetitive in a company and it uses a relatively sizeable project team, the project manager delegates the operational running of the process to a member of the company's quality department who is made available to him. When this is not the case, and the project manager is often also the project promoter, he must control the operational running of this process. He may gather advice when the project is nurtured in an "incubator", or, within a company, use the specific departments in charge of relations with the project's various stakeholders (human resources for workers, financial department for investors, quality-environment-safety department for regulatory authorities, marketing department for clients, etc.).

The managerial organization of the process can be described with the help of the following model:

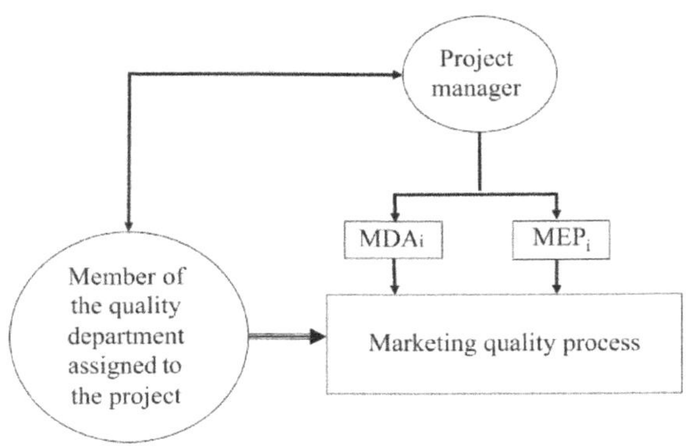

Figure 4.1. *Managerial organization of a "marketing quality" processor*

COMMENT ON FIGURE 4.1.–

– MDAi denotes the members of the management within the company responsible for relations with a specific class of stakeholders (human resources management, financial service management, marketing management, purchasing management, administrative management, legal department, etc.).

– MEPi denotes the members of project team number (i) who contribute to the added value of the innovation process by supplying a particular type of skills related to an area of expertise.

Relations between project manager and other members of the processor are *managerial relations by delegation*. The various members who participate alongside the project manager are seconded to the project. They belong to other "managerial areas" and made available for the sake of the project. The project manager gives orders to these members and provides the resources for them to carry such orders out. These relations must take place in a strong spirit of partnership, whereby the project manager and these members must work together in accordance with the order statement, the definition of the resources to be assembled, and these members can request assistance from the project manager in carrying out such tasks.

The initial meetings are generally dedicated to ensuring that the processor is able to produce the marketing quality of the four processes which guarantee the production and use of quality in the innovation process. The project manager ensures that the members of this processor share the opportunity and feasibility of such a process. He also validates the specific scenario regarding a given process by distributing roles among the various team members.

The project manager then decides to initiate the process of designing scenarios for the four processes for the production and use of quality.

As before, for the production of CQT, throughout the entire process, the project manager oversees the group's ability to collectively produce the desired results, polices any deviations, checks the relevance of information released and ensures that decisions are shared.

He requests that the quality department propose a system for assessing the risks of not reaching the processor's objectives and to carry out the necessary alerts.

He makes decisions on corrective action, depending on the problems observed.

Throughout the process, the project manager must show the interest he attaches to the design of marketing quality scenarios, through his behavior and attitude. He must show that the information released will produce real added value to give rise to competitiveness factors. He must also show that he is ready to overcome the challenges which have emerged. It is essential that the motivation of the processor's human resources is maintained at a high level throughout the process.

At the end of the process, the project manager is to carry out a report on the initial relevance of the scenarios, the level of consensus in the choices made and the ability of team members to then work together in complying with these scenarios.

4.7. Conclusion

The key stage of the CQP integrated into an innovation process is often dealt with too quickly because it does not contribute to the production of added value for the innovation process, but to the production and use of this added value's quality.

This type of action, in an innovation process, is often seen as a brake on unbridled production of creative research which is perceived as necessary for finding an original solution to a need of the market.

It is not a matter of waiting for the various scenarios to be written out perfectly and validated before starting to produce this added value. These scenarios can be designed step by step as previously indicated in the section on planning these actions.

This stage is also difficult because it requires team members, who need to concentrate on their ideas in order to capitalize on them fully, to anticipate the behavior and attitudes of other members whose contributions are essential for the project to be successful. They must reflect on how best to satisfy them and make the most of the potential of these contributions, while they are already wrestling with the problem of organizing their own ideas, even though these two processes are not based on the same logical approaches. These two processes cannot be carried out simultaneously.

There are several reasons for including the concept of marketing quality in an innovation process. It only regards certain events which can be easily identified in the four key processes for the company's production and use of quality. Factual descriptions are given for the scenarios of events which seem the most realistic and likely to contribute to achieving the desired results, without dealing with how they will be produced. By focusing on these four processes in this way, the field of action is clearly specified for the quality process integrated into an innovation process. Targeting factual events makes the role of quality more tangible. Finally, separating events in a process and the configuration of the processor which will cause them avoids confusion or mix-ups and reduces the problems in designing systems for the production and use of quality within an innovation process.

For the implementation of this stage of a CQP integrated into an innovation process, like with all the other stages to follow, it is necessary to distinguish between two situations:

– innovation processes which are repetitive and concentrated in a specialized department of the company;

– the innovation process is a single project which will give rise to a company responsible for its economic development.

In the latter case, relations with stakeholders other than future clients of the products incorporating the innovation are relatively standardized and can be reproduced. The marketing quality of the four processes for the production and use of CQF is based on the scenarios, the bulk of which will have already been written. The same can be said for the other stages of the quality process.

To take account of new CQTs, it is simply necessary to revise these stages in order to adjust certain specifications.

In the former case, the situation is completely different. Everything must be discovered and created. This situation is more complex, but it leaves a greater amount of freedom. This is why significant ground-breaking innovations often come from start-ups, and this is also why large companies often look for innovations among start-up companies.

5

The "Quality" Fuel of an Innovation

5.1. Introduction

The design of an innovation's competitive quality fuel is one of the most original aspects of the quality process integrated into an innovation process.

The ideas addressed in this chapter are largely put into practice by companies which have relied on competitive quality for a number of years in order to ensure the success of their innovation projects. However, since they try to keep this aspect of their operations as confidential as possible, they often use a vocabulary which is specific to their individual companies. In order to present the general models used, we decided to introduce our own terms which have already been extensively used in other published works [MAI 11, MAI 13].

We defined an innovation's *competitive quality fuel* (CQF) for a specific class of stakeholders as:

> *The collection of all the information released by the processor and innovation process, its results, and the effects they produce, which have an impact on the competitive quality judgments of the class of stakeholders.*

The design of this quality fuel is affected by the predefined competitive quality tactic (CQT) and by the marketing quality of the four quality production and use processes within the innovation process.

During this design stage, new ideas and information which arise can lead to a modification of the CQT and the marketing quality of the four processes.

It is important to recall the following definition of the competitive quality process' (CQP) technical function, the study of which is the subject of this chapter:

> *An innovation's CQF are to be designed so as to maximize the chances of the competitive quality tactics being fulfilled with the minimum amount of energy being expended.*

It is the design of the information which should stimulate the stakeholders' quality perceptions, with regard to marketing quality scenarios, in order to prompt their competitive quality judgments encouraging them to provide the compensatory resources or services expected in compliance with the provisions in the CQT.

This information can be released by the processor which carries out the innovation process, certain events in the innovation process, various configurations of the innovation as it progresses, certain results of impact and feasibility studies, and uses of the innovation which arise during the process.

This chapter is structured in the same way as the previous two chapters.

5.2. An innovation process' competitive quality fuel

We examined earlier how a competitive quality judgment prompting a stakeholder to provide the expected compensatory resources is the result of given stakeholder perceptions.

They were defined in Chapter 2 as follows:

– perceptions of expectancy regarding:

– the satisfaction of expectations by certain components of the innovation production process, its results or the effects it causes;

– the ability to use these components;

– the ability to make the work required to understand, obtain, adjust and use them profitable;

– perceptions of confidence in the ability of the innovation process to comply with the objectives given.

For each class of stakeholders, the CQTs aimed at the specific competitive quality perceptions to be stimulated, and the marketing quality provides a description of certain events to be enacted particularly during the release of CQF, and during the emergence of these perceptions to prompt competitive quality judgments.

The design of CQF regarding one class of stakeholders consists of identifying and characterizing the information to be released:

– by activities of the innovation process;

– by characteristics of the processor which carries them out;

– by evolutions in the configuration of the innovation itself;

– by the effects which it is likely to cause in its environment in order to stimulate the competitive quality perceptions targeted by the CQT in the best way possible.

This work must be completed for each class of stakeholders in the innovation process targeted by the CQT.

Some examples

Let us suppose that an innovation regards the creation of a production system for made-to-measure artificial hip joints which are adapted to the physiology of each patient. One of the stakeholders targeted by the CQT is always the French National Health Authority (FNHA) which is responsible for evaluating whether it offers an appropriate response to the need for services required by patients and medical teams, and to help to establish the amount it should be reimbursed by the Social Security system.

One of the first pieces of information required by the National Health Authority is a report produced by a registered body, which certifies that each made-to-measure joint has "CE" marking. Obtaining this certification means that the unitary production system for the joints is under sufficiently close control for it to be possible to certify that each joint will satisfy the specification of CE marking.

This information prompts the FNHA's competitive quality perceptions because it satisfies one of the specifications, which is the credible demonstration of the ability of the future production system to produce joints compliant with the performance claimed. It gives confidence in the response to this type of specification because it is released by a registered body which is accredited to award it CE marking. The FNHA can use this information to partially respond to evaluation criteria regarding the safety of the patients having these joints fitted. This information is therefore part of the innovation's CQF.

However, this information does not show whether the production of made-to-measure artificial hip joints constitutes real medical progress. Does the patient benefit from the joint, does it offer advantages compared with the best traditional artificial hip joints currently available on the market?

Only significant clinical trials can help the FNHA to answer this question. This new category of information may or may not satisfy the FNHA specifications. These are contained within a set of specifications which specify how these clinical trials must be carried out so that the results can be accepted. The content of the report on these clinical trials is also part of the information which makes up the CQF of the innovation.

In a different field, the innovative style of a car also makes up part of its CQF.

The innovative mechanical complexity of a luxury watch is another example of a product's CQF.

Systematically causing a rise in the price of watches, in auction rooms, in order to ensure that the second-hand price is higher than the initial sale price is a piece of information which also makes up part of the CQF when the class of stakeholders is the market.

5.3. Planning the design of competitive quality fuel in the innovation process

The following box shows the stages in the innovation process where the CQF of these various classes of stakeholders must be designed.

Category of stakeholders	Stages in the innovation process
Future users of products or services incorporating the innovation	– Creating experimental drafts which provide a more accurate representation of the idea; – Conducting prospective studies on possible socioeconomic uses for the innovation, and its impacts on the company and its position; – Progressively designing drafts to accurately represent the configuration of the innovation; – Progressively designing drafts which give a representation of future products and services to incorporate the innovation, by describing, in particular, the future "interfaces" of the innovation with other components of these products and services; – Characterizing future uses for products or services which would incorporate the innovation; – Conducting performance tests on future products or services.
Potential investors in the innovation and other stakeholders contributing to the project	– Conducting initial feasibility tests on the future products or services; – Conducting prospective studies on possible socioeconomic uses for the innovation, and its impacts on the company and its position; – Briefly characterizing products or services which could benefit from the innovation; – Progressively designing drafts which give a representation of future products and services to incorporate the innovation, by describing, in particular, the future "interfaces" of the innovation with other components of these products and services; – Conducting tests on the drafts of future products or services, at various steps in their production, on future clients and on all future stakeholders in the production system for the products and services; – Conducting studies on the nature and feasibility of exchanges with future stakeholders in the project necessary for it to be fully realized; – Conducting performance tests on future products or services;

	– Identifying stakeholders who will contribute to the development and production of the future products and services which will incorporate the innovation; – Building a processor responsible for carrying out the innovation process and an external communication network.
Future contributors to the "business unit" which will produce and distribute the products or services incorporating the innovation	– Conducting prospective studies on possible socioeconomic uses for the innovation, and its impacts on the company and its position; – Identifying stakeholders who will contribute to the development and production of the future products and services which will incorporate the innovation; – Designing prototypes of the production and distribution systems for the products and services; – Conducting studies on the ability of the company to design production and distribution systems for the products and services which will incorporate the innovation by taking into account the configurations of the various definitive drafts of the innovation; – Conducting tests on the drafts of future products or services, at various steps in their production, on future clients and on all future stakeholders in the production system for the products and services; – Characterizing future uses for products or services which would incorporate the innovation; – Conducting performance tests on future products or services; – Examining the steps to take to ensure the intellectual protection of the innovation.

Box 5.1. *Stages of an innovation process where the CQF of an innovation is designed*

5.4. Ways of defining competitive quality fuel

The design of CQF means predicting how certain characteristics of the information released can have an effect on stakeholders' competitive quality perceptions when compliant with marketing quality scenarios.

This design process generally consists of the following stages:

– recalling CQT data regarding the competitive quality perceptions to be prompted for a given class of stakeholders;

– ascertaining the nature of the information which may be released to prompt these perceptions during various stages of the innovation process;

– placing the release of this information into the scenarios of competitive quality marketing;

– locating the various vehicles possible for this information by identifying sub-processors of the main processor carrying out the innovation process, the events of the innovation process, components of the innovation's configuration and the effects that it may produce on its environment which may release this information;

– designing the specifications of each piece of information so that they have the best chance of prompting these perceptions when released;

– validating these specifications by using various drafts of the innovation, and future production and distribution systems of the products or services which will incorporate it, predictions on future uses of the products or services, as well as carrying out performance tests on future products or services.

The design of the information specifications for this CQF essentially addresses how these specifications can act on the operation methods of the stakeholders' quality oxidizer to obtain the desired impact on their competitive quality perceptions.

These specifications are sometimes demands which are formulated by stakeholders. In this case, they simply need to be checked.

However, in general, an innovation process means predicting an unknown future which requires constant revisions of the definition of these specifications according to new opportunities or related constraints which emerge during the process. These specifications are not defined in order to systematically ensure compliance throughout the process, but to provide temporary collective guidelines which are defined by project team members, through their language and logical approach.

These guidelines channel the project team members' collective action and make up part of the backbone of the project. They are one of the interfaces between the project and its stakeholders which structure and rationalize the exchanges without reducing the freedom necessary for creating added value through creativity.

These guidelines apply to two of the three lines of action which are encountered during exchanges with stakeholders.

The three lines of action can be described in the following way:

– standard rules which systematically apply regardless of the situation;

– conditional "if ... then ..."-type rules which depend on various predefined situations;

– lack of rules compensated by an individual competence which can be used to adapt to and profit from unexpected situations.

The CQF specifications only apply to the first two situations.

Other sources for producing these specifications arise from using common sense and methods of creativity, capitalizing on experience, observing phenomena which seem to have no link to the subject in question, but where similarities could provide ideas.

Similar methods which are commonly encountered in the process of designing products and service known as "quality function deployment" (QFD) can help to fulfill this stage.

These matrices give a simple, quantitative and systemic representation of successive cause–effect links between these information specifications, how to obtain them, then interpret them in terms of quality perceptions and, finally, rank them according to the significance of the various perceptions attached to each one.

These methods are still not being used well. Often, analysis is carried out with too much detail accumulating a large amount of information, meaning that the use of this message becomes too complicated. They must be targeted on the impacts on stakeholders' competitive quality perceptions which are the most significant and the hardest to anticipate.

The stages of design for a competitive quality fuel can be summarized as follows:

– recall the target of the CQF information as defined by the CQT;

– identify information likely to be included;

– include planning for the release of information in the competitive marketing quality scenarios;

– identify the emitters of this information;

– define the specifications of this information;

– validate these specification through simulations.

Box 5.2. *Stages involved in the design of competitive quality fuel*

In the example of artificial hip joints, the stakeholder to be persuaded is the National Health Authority. The competitive quality perceptions targeted by the CQT which need to be acted upon are perceptions of:

– an original response to expectations of a significant medical improvement in the field of artificial hip joints;

– a system for organizing the availability of artificial joints for medical teams compatible with their operating methods;

– real medical cost-effectiveness of these changes to operating practices necessary for transplanting these joints;

– enough confidence in the ability of the innovation process, and the innovation itself, to meet the objectives established.

Depending on these perceptions, the *Haute Autorité de la Santé* (HAS) authorizes the entry into the market of this innovation, with an estimate of the significance of the improvement to the medical service which can be expected.

Three types of information should be prioritized in order to prompt these perceptions:

– the certification by a registered body on the compliance with regulations allowing CE marking of these joints;

– the reports on clinical trials proving the performance of the new joints showing its superiority in comparison to competitors;

– demonstration of the ability of delivering the made-to-measure joint on time and without errors so as not to disrupt the care process.

The design of these information specifications is guided by using European regulation for the attribution of CE marking in terms of operational demands which can be applied to the company's activities, and in terms of the rules imposed by the NHA on the demonstration of the innovation's performance offered by clinical trials. This design must also take into account the model of the production and delivery supply chain for these joints, as well as its control method.

In a general way, determining the nature of the information which can be released to prompt competitive quality perceptions is largely guided by the prior marketing quality scenarios.

Designing the information specifications is complex since they must take into account a large number of constraints. This stage is often carried out with an empirical approach.

This design must take into account the known characteristics of the stakeholders' quality comburant and the relative importance that they attribute to each category of quality perception.

It is necessary to clarify how the stakeholders may obtain, interpret and evaluate the CQF information of the products or services to prompt their competitive quality perceptions. However, part of the stakeholder group is particularly built on the basis of characteristics common to their CQF. When the stakeholders are clients, it is the responsibility of the marketing department to establish the general operational mechanisms of clients' CQF.

Designing the information specifications for the CQF is mainly based on how they may impact the operation of stakeholders' CQF in order to achieve the desired effect on their competitive quality perceptions.

The previous stages of the CQP generally provide the information which guides the studies intended to establish the operation of stakeholders' quality comburants. Observing how stakeholders operate in their general processes of consumption of products or services helps to highlight the main characteristics of their quality comburants. When possible, it is also useful to observe how stakeholders operate when they receive information from competitors. These characteristics can be supplemented by interpreting certain signals which are released, or certain events produced by the behavior or attitude of stakeholders, during situations other than those caused by consumption processes. It is possible to make mistakes when developing these interpretations. Well targeted tests can be used to confirm, or contradict, a given interpretation.

For example, the way a client questions the origin of certain parts of the product, or how they have been made, can provide information on aspects of its technical expertise which will influence its interpretation of a product's quality fuel, and prompt some of its competitive quality perceptions.

There are a number of methods for obtaining an overview of the stakeholders in order to identify behavior which shows how their quality comburants acts on their perceptions. Exchanges between "front office" staff and stakeholders help to further highlight these signs. Please note: a questionnaire is rarely able to correctly capture this type of information. Scientific studies show that the bias introduced by this type of questionnaire significantly masks the reality of quality comburants' operations.

It is essential to identify the stakeholder's quality comburant. For instance, an innovation may be able to satisfy a stakeholder's need which is as yet unanswered, but it may not be driven to obtain the product or service incorporating this innovation because its quality comburant provides a negative interpretation of some of the information released by the company, product or service. For example, when an innovation appears to significantly change clients' consumption habits and would entail a long-winded learning process, clients will not buy the product incorporating the innovation, even if it could satisfy one of their clearest needs.

COMMENT.– In general, companies are familiar with this type of process when stakeholders are clients. In the current situation, we will generalize these processes to all classes of stakeholders.

5.5. Performance of the design of competitive quality fuel

The first stage for evaluating the performance of this design consists of ensuring that the scenarios of the process for releasing CQF allow the release of previously developed information in compliance with their specifications. To do this, the events described by these scenarios need to be reviewed and it must be verified that the information specified by the CQF design will be correctly released during the process.

For example, if a specific characteristic regarding the shape of the product is a piece of information which is part of its competitive quality fuel, the company must ensure that certain key events in the scenario of the process for releasing this CQF highlight this characteristic, by drawing stakeholders' attention to it.

If this characteristic is the color of a product sold in a supermarket for general consumption, it is necessary to ensure that the positioning of this product in a supermarket's display clearly shows this color.

For the second stage of evaluation, two situations can be distinguished:

– either the stakeholders express precise expectations or even demands regarding the specification of certain aspects of information for the CQF;

– or innovation attempts to respond to certain needs that the stakeholder struggles to express, and which become more precise as the functionalities of the innovation are revealed.

In the first case, measures of performance are limited to traditional compliance measurements. They can start as soon as the design stage of the information specifications is completed. It continues to the stage dealing with the validation of these specifications.

In the second case, the stages regarding the design of CQF, as described in the previous paragraph, show that it is difficult to measure the performance of this process before designing drafts of the information emitters. During the tests of these emitters, which consist of surveying the stakeholders in question, the analysis of their behavioral reactions helps to estimate a performance level for this design process.

The third evaluation stage takes place when the stakeholders actually obtain the information regarding the CQF of the innovation targeted at them. It is a real-time evaluation which is carried out by observing the attitude and behavior of stakeholders when they identify this information, and possibly give their competitive quality judgment. It is important at this stage to verify if the CQF released would allow for a correct fulfillment of the defined CQTs.

The fourth evaluation stage comes when the company directors carry out an interim or final analysis of the benefits achieved by applying the CQTs. During this analysis, the company attempts to identify the real contribution of various CQFs of the innovation in obtaining these benefits. This final evaluation stage helps to provide ideas for the improvement processes.

5.6. Managing the design of competitive quality fuel

Designing competitive quality fuel for a specific class of stakeholders is a process placed under the responsibility of the project manager, who will lead the process.

Carrying out this process consists of defining specifications for certain pieces of information which will be released during the innovation process. As before, it uses specific expertise which is enhanced thanks to feedback.

When the company repeats an innovation process and uses a relatively large project team, the project manager delegates the operation of the process of defining CQF specification to a member of his team who is part of the company's department responsible for relations with the class of stakeholders in question (human resources, purchasing, finance, marketing, etc.).

A member of the company's quality department is made available to assist in carrying out this process and to ensure links with other technical functions in the CQP integrated into an innovation process.

When this is not the case, and the project manager is often also the project promoter, he must control the operational running of this process. He may gather advice when the project is nurtured in an "incubator", or, within a company, use the specific departments in charge of relations with the project's various stakeholders.

The managerial organization of the process is shown in Figure 5.1.

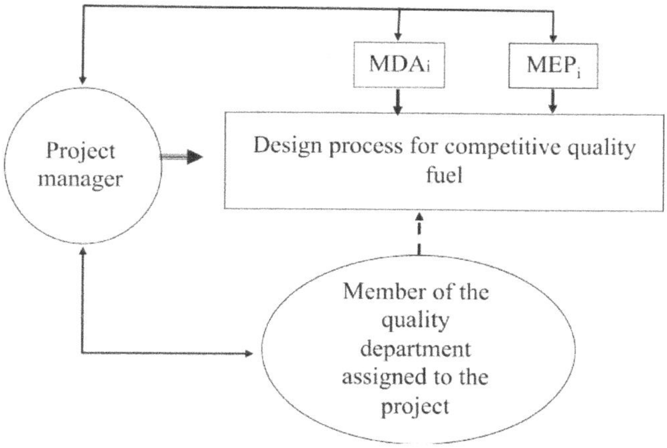

Figure 5.1. *Managerial organization of a processor for designing competitive quality fuel*

COMMENT ON FIGURE 5.1.–

– MDA_i denotes the members of the management within the company responsible for relations with a specific class of stakeholders (human resources management, financial service management, marketing management, purchasing management, administrative management, legal department, etc.).

– MEP_i denotes the members of project team number (i) who contribute to the added value of the innovation process by supplying a particular type of skills related to an area of expertise.

Relations between project manager and other members of the processor are, as before, managerial relations by delegation. The various members who participate alongside the project manager are seconded to the project. They belong to other "managerial areas" and made available for the sake of the project. The project manager gives orders to these members and provides the resources for them to carry such orders out. These relations must take place in a strong spirit of partnership, whereby the project manager and these members must work together in accordance with the order statement, the definition of the resources to be assembled, and these members can request assistance from the project manager in carrying out the project.

The initial meetings are generally dedicated to ensuring that the processor is able to produce the CQF of the innovation. The project manager ensures that the members of this processor share the opportunity and feasibility of such a process. He also validates the specific scenario regarding a given process by distributing roles among the various team members.

As before, throughout the entire process, the project manager oversees the group's ability to collectively produce the desired results, polices any deviations, checks the relevance of information released and ensures that decisions are shared.

He requests that the quality department propose a system for assessing the risks of not reaching the processor's objectives and to carry out the necessary alerts.

He makes decisions on corrective action, depending on the problems observed.

Throughout the process, the project manager must show the interest he attaches to the design of this CQF, through his behavior and attitude. He must show that the information released will produce real added value to give rise to competitiveness factors. He must also show that he is ready to overcome the challenges which have emerged. It is essential that the motivation of the processor's human resources is maintained at a high level throughout the process.

At the end of the process, the project manager needs to carry out a report on the initial relevance of the specifications attributed to the information which makes up the CQF, the level of consensus in the choices made and the ability of team members to then work together in complying with the production of these specifications.

5.7. Conclusion

This essential stage of the quality process integrated into an innovation process must leave a significant amount of room open for creativity. The only constraints are the definition of the CQT, and the scenarios of the marketing quality process. However, it is more a set of guidelines for directing the team's action than constraints. Indeed, the design of the specifications for the information that will make up the CQF can give rise to opportunities or risks which lead to a modification of the marketing quality results, or the directions on the CQTs.

It is very important to observe that the design of these specifications for information does not affect the configuration of the future processors which will be responsible for producing and releasing this information.

While it is always preferable to examine the feasibility of the specifications designed in this way, a localization or identification of processors which could produce or release this information must leave a large area of freedom for their future design. This is the role of the following stage in the CQP.

A segmentation of these tasks is necessary so as not to harm the creative performance of the project members.

Finally, this stage must draw upon the collaborative work of the various members of the project team.

6

The Emitters of Quality Fuel

6.1. Introduction

The emitters of competitive quality fuel (CQF) for various classes of stakeholders in the innovation process are targeted by the competitive quality tactic (CQT), by marketing quality scenarios and, finally, by the definition of CQF specifications.

The previous stages in the competitive quality process (CQP) help to focus energies within the project on precise and limited objectives in order to spread the information which prompts stakeholders' competitive quality judgments to ensure the success of the innovation.

The design of these emitters does not always constitute an additional investment. It has often been observed that they have uses other than for releasing CQF. If so, it is simply a case of checking and occasionally modifying their functionalities or technical specifications. When they do not directly regard the production of the innovation, but supplementary products or services for stakeholders other than the future users of those incorporating the innovation, these emitters are often the same as those which were already used for other innovation projects. Once again, it is simply a case of adjusting their configuration so that they can release CQF in compliance with their specifications.

For instance, a business plan makes up parts of the CQF emitters for project investors. Its configuration varies little from one stakeholder to another. However, the primary information which will be transformed to obtain the CQF in compliance with their specification is specific to each project.

The aim of this chapter is to offer general guidelines for these characterizations, evolutions or conceptions.

Its structure is the same as the previous three chapters.

6.2. Competitive quality fuel emitters

CQF emitters are systems which produce the information to make up the CQF and comply with the predefined specifications. They release the information so it can be easily obtained by the targeted stakeholder and interpreted by its quality oxidizer in order to produce the competitive quality judgments defined by the CQT.

The release of this information is planned by the competitive marketing quality.

These emitters are processors, as defined in section 1.2.2 of Chapter 1.

Their activities are directed by a managerial organization specific to each emitter.

They are designed once for the entire duration of the project. They are activated by sequences of released information which comply with the project's marketing quality scenarios.

The information which forms the basis of these emitters is also known as "primary information" and is generated by the innovation process. At its heart lies the evolution of the configuration of the innovation's functional potential. This information feeds the CQF of future users of the products or services incorporating the innovation, but also investors, workers in the project team, and all the other stakeholders who make up a part of the project's socioeconomic driver.

For example, communication regarding a component of a new medicine's functional potential, which is likely to have an impact on the state of a pathology according to some indications, makes up part of the CQF's emitters for doctors, patients, investors, bodies responsible for public health and so on.

It is occasionally desirable to design different emitters to describe the primary information which characterizes this functional potential when their information is part of the CQF for various classes of stakeholders. Evidently, the same information will not be communicated in the same way to future clients, future investors or future regulating bodies.

At this stage in the CQP, the information has already been specified for each class of stakeholders. Drivers now need to be designed, which will create the primary information, transform it into secondary information in compliance with the CQF specifications and send it to the stakeholders targeted.

The possible effects of the innovation on the socioeconomic environment are not changes in the state of its functional potential, but the effects that these changes may cause on this environment as they trickle down. For instance, innovations in medical imaging have significantly changed the distribution of roles between radiologists and surgeons in treating certain illnesses. Communication regarding these predictable effects can be part of the information in the CQF for certain stakeholders. In general, a large section of these stakeholders are concerned by these effects. Once again, it may be necessary to personalize the way that this information describing the effects is communicated.

Information about certain events in the innovation process which are not direct factors in the evolution of the innovation's functional potential may be part of some CQF. For instance, the way in which the performances of an innovation's new functionalities are tested may have an effect on the competitive quality perceptions of several classes of stakeholders. The way in which this information is communicated then becomes essential.

In the same way, certain characteristics of the processor which carries out the innovation process may be information which forms part of some CQF. For example, the skills of certain members of the project team, the charisma of the project manager, the reputation of certain providers or investors in the project, and the performance of some of the project's material resources can have an effect on other stakeholders' competitive quality perceptions.

Finally, independently of the innovation itself, the specific products or services offered by the company to one of the classes of stakeholders release information which may be a component of their CQF. For instance, bonuses awarded to members of the project team depending on the success of the innovation may stimulate their interest in contributing to the project.

6.3. Defining a competitive quality fuel emitter

The following stages of the CQP define:

– the specifications of the information to be released which forms the CQF;

– events in the project's products or services, possibly co-produced with other stakeholders, and during which information is released;

– a predefined and organized flow diagram of these events;

– the events likely to be observed which describe how the information in the CQF have an impact on the stakeholders' competitive quality perceptions;

– a probable flow diagram of these events;

– the various possible vehicles for this information located in the processors which carry out the innovation process, formalizing and communicating the various stages in the evolution of the innovation's configuration, and the effects that it is likely to produce on its socioeconomic environment, or which are specific to the production of products or services offered to each class of stakeholders.

This is input data in the design process for CQF emitters, which evolve as the process progresses. As a consequence, the project manager must frequently update the way that the CQF emitters are activated. This may mean modifying their configuration to adapt them to these evolutions.

The emitters are located in the following systems:

– the processor carrying out the innovation process;

– the processor formalizing and communicating the various stages in the evolution of the innovation's configuration;

– the processors carrying out certain technical functions in the CQP;

– the processor formalizing and communicating the effects that these stages are likely to produce on the company's socioeconomic environment;

– the specific processors for the production of the products or services offered to each class of stakeholders.

Depending on the type of information, the emitters fulfill different functions:

– communication of raw data produced by the first four processors (Type 1) (*be aware of the confidentiality of the information communicated*);

– communication of filtered information, produced by the first four processors, which could prompt a negative competitive quality perception from some stakeholders (Type 2);

– transformation of some information produced by the first four processors, which are not aimed specifically at certain stakeholders, but which could prompt positive competitive quality perceptions (Type 3) (*innovation selling points*);

– communication of information produced specifically for certain products or services offered to a class of stakeholders without explicitly describing the progress of the innovation process (Type 4) (*personalization of products or services*).

The emitters which fulfill type 1 functions must be designed with a great deal of caution because they may reveal certain original components of the future innovation which could be used by competitors. Nevertheless, this information is often necessary in order to attract certain stakeholders. They must be made to sign a confidentiality contract, or limit the precision of the information given to make it impossible for competitors to use it for their own advantage. These emitters must only be activated by the project manager. (Please note: during an audit process carried out by a registered body, this information must not be communicated.)

For instance, the presentation of a "concept car" at a car show forms part of these emitters.

The emitters which fulfill type 2 functions must filter information used to help the progress of the innovation process, but which could unconstructively disrupt relations with certain stakeholders. The complexity of the connections in the driver of socioeconomic exchanges of an innovation process should promote the need to control any rumors based on intentional or unintentional information leaks. Often, releasing a summarized and highly abstract version of this information helps to fulfill this function.

For example, the failure of negotiations with some potential stakeholders may be considered by others as a sign of a risk of failure in the innovation process, and must remain confidential when it can be compensated by other on-going negotiations.

The emitters which fulfill type 3 functions must give structure to that information which does not directly concern the products or services targeted at a class of stakeholders, but which can have a positive effect on some of their competitive quality perceptions to encourage them to provide the expected contributions more easily. These are often known as "CQF amplifiers".

Emitters which fulfill type 4 functions must:

– personalize the communication of information regarding how the products or services offered to stakeholders can satisfy their expectations through their use;

– facilitate the access of stakeholders to all the information which constitutes the CQF of these products or services;

– help stakeholders understand this information;

– emphasize the signs of credibility for this information.

The processes of releasing information which fulfill type 4 functions are often placed in the category of "service process" process. This means that they are systems of co-production between certain members of the project team and the stakeholders in question. This system of co-production is always led by the project manager.

Type 4 functions generally answer specific questions from a class of stakeholders. These answers are not directly produced at the center of the innovation process. (For example, business plan for investors, bonuses for members of the project team, pay systems for providers and manually ensuring the safety of the products incorporating the innovation).

The emitters which fulfill these functions are located in the processors which produce the particular components for the products or services in question.

The project team is often tempted to neglect the quality of these components because they do not directly contribute to progressing the innovation's configuration which lies at the heart of the production of added value.

The design of these emitters is a process which generally consists of the following stages:

– spatial and temporal localization of the five processors identified above, and possible productions of primary information which could be used to release future CQF;

– highlighting the characteristics of this primary information;

– designing scenarios in the process of transforming this primary information into CQF which complies with their specifications;

– design the processor which will obtain the primary information, transform it into CQF and communicate the CQF to stakeholders.

One or several emitters must be allocated to each information source. These emitters are individual processors and their design consists of specifying the characteristics of their configuration.

It should be recalled that the "configuration" of a system is made up of the structure of its resources, which may be human, material, information-based, organizational and financial, along with its functional potential, which is the collection of its functionalities and the time-based organization of its actions (process).

The definition of these specifications is fully integrated into the various actions of the innovation process' production.

It is therefore impossible to offer a general method for creating definitions for these specifications. However, the project manager must ensure that the design of these emitters is carried out at the correct time.

The following box provides guidelines for directing the planning of these definitions.

6.4. Planning the design of competitive quality fuel emitters

The box showing the stages in the innovation process where the specifications for CQF emitters for various classes of stakeholder must be established is identical to that described in the previous chapter.

Category of stakeholders	Stages in the innovation process
Future clients of the products and services incorporating the innovation	– Conducting prospective studies on possible socioeconomic uses for the innovation, and its impacts on the company and its position; – Conducting studies on the nature and feasibility of exchanges with future stakeholders in the project necessary for it to be fully realized; – Progressively designing drafts to accurately represent the configuration of the innovation; – Progressively designing drafts which give a representation of future products and services to incorporate the innovation, by describing, in particular, the future "interfaces" of the innovation with other components of these products and services; – Conducting performance tests on future products or services; – Characterizing future uses for products or services which would incorporate the innovation; – Designing prototypes of the production and distribution systems for the products and services; – Conducting studies on the ability of the company to design production and distribution systems for the products and services which will incorporate the innovation by taking into account the configurations of the various definitive drafts of the innovation.
Potential investors in the innovation and other stakeholders who contribute to the project	– Identifying possible clients for the future products or services likely to incorporate the innovation; – Creating experimental drafts which provide a more accurate representation of the idea; – Conducting prospective studies on possible socioeconomic uses for the innovation, and its impacts on the company and its position; – Building a system responsible for carrying out the innovation process and an external communication network;

	– Progressively designing drafts to accurately represent the configuration of the innovation; – Progressively designing drafts which give a representation of future products and services to incorporate the innovation, by describing, in particular, the future "interfaces" of the innovation with other components of these products and services; – Conducting performance tests on future products or services; – Characterizing future uses for products or services which would incorporate the innovation; – Conducting studies on the ability of the company to design production and distribution systems for the products and services which will incorporate the innovation by taking into account the configurations of the various definitive drafts of the innovation; – Conducting tests on the drafts of future products or services, at various steps in their production, on future clients and on all future stakeholders in the production system for the products and services; – Creating a rationale to aid decision-making which examines the probable socioeconomic impacts of implementing the innovation; – Creating a rationale to aid decision-making which details the feasibility of implementing the innovation.
Future contributors to the business unit which will produce and distribute the products or services incorporating the innovation	– Identifying possible clients for the future products or services likely to incorporate the innovation; – Conducting prospective studies on possible socioeconomic uses for the innovation, and its impacts on the company and its position; – Progressively designing drafts which give a representation of future products and services to incorporate the innovation, by describing, in particular, the future "interfaces" of the innovation with other components of these products and services;

	– Conducting performance tests on future products or services; – Characterizing future uses for products or services which would incorporate the innovation; – Identifying stakeholders who will contribute to the development and production of the future products and services which will incorporate the innovation; – Designing prototypes of the production and distribution systems for the products and services; – Conducting studies on the ability of the company to design production and distribution systems for the products and services which will incorporate the innovation by taking into account the configurations of the various definitive drafts of the innovation; – Conducting tests on the drafts of future products or services, at various steps in their production, on future clients and on all future stakeholders in the production system for the products and services; – Creating a rationale to aid decision-making which examines the probable socioeconomic impacts of implementing the innovation; – Creating a rationale to aid decision-making which details the feasibility of implementing the innovation; – Examining the steps to take to ensure the intellectual protection of the innovation.

Box 6.1. *Stages of the innovation process where competitive quality fuel emitters are created*

6.5. The performance of the design of competitive quality fuel emitters

Measuring the performance of the emitters' design is limited to:

– a measurement of the compliance between the information released and the CQF specifications;

– compliance with plans establishing the progress of events for the production of CQF.

6.6. Managing the design of quality fuel emitters

As in the previous chapter, designing a CQF emitter is a process placed under the responsibility of the project manager who will lead the process.

Carrying out this process consists of defining specifications for CQF emitters which give rise to this information. As before, this procedure uses specific expertise which is enhanced due to feedback.

When the company repeats an innovation process and uses a relatively large project team, the project manager delegates the operation of the process of defining CQF emitter specifications to a member of his team who is part of the company's department responsible for relations with the class of stakeholders in question (human resources, purchasing, finance, marketing, etc.).

A member of the company's quality department is made available to assist in carrying out this process and to ensure links with other technical functions in the CQP integrated into an innovation process.

When this is not the case, and the project manager is often also the project promoter, he must control the operational running of this process. He may gather advice when the project is nurtured in an "incubator", or, within a company, use the specific departments in charge of relations with the project's various stakeholders.

The managerial organization of the process can be described with the help of the following model:

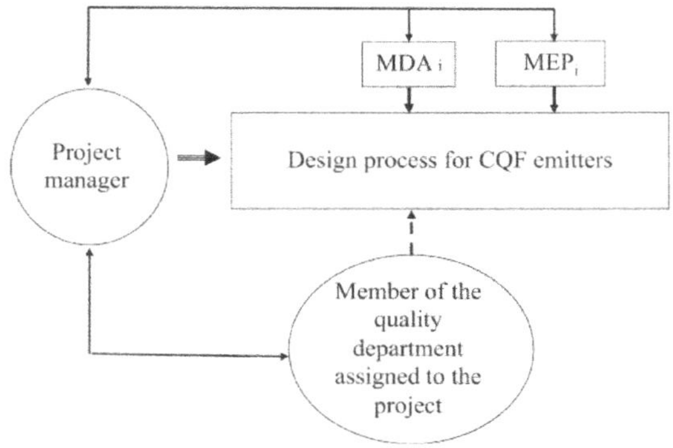

Figure 6.1. *Managerial organization of a processor for the design of quality fuel emitters*

COMMENT ON FIGURE 6.1.–

– MDA_i denotes the members of the management within the company responsible for relations with a specific class of stakeholders (human resources management, financial service management, marketing management, purchasing management, administrative management, legal department, etc.).

– MEP_i denotes the members of project team number (i) who contribute to the added value of the innovation process by supplying a particular type of skills related to an area of expertise.

Relations between the project manager and other members of the processor are, as before, managerial relations by delegation. The various members who participate alongside the project manager are seconded to the project. They belong to other "managerial areas" and made available for the sake of the project. The project manager gives orders to these members and provides the resources for them to carry such orders out. These relations must take place in a strong spirit of partnership, whereby the project manager and these members must work together in accordance with the order statement, the definition of the resources to be assembled, and these members can request assistance from the project manager in carrying out the project.

The initial meetings are generally dedicated to ensuring that the processor is able to produce the CQF emitters for the innovation. The project manager ensures that the members of this processor share the opportunity and feasibility of such a process. He also validates the specific scenario regarding a given process by distributing roles among the various team members.

As before, throughout the entire process, the project manager oversees the group's ability to collectively produce the desired results, polices any deviations, checks the relevance of information released and ensures that decisions are shared.

He requests that the quality department propose a system for assessing the risks of not reaching the processor's objectives and to carry out the necessary alerts.

He makes decisions on corrective action, depending on the problems observed.

Throughout the process, the project manager must show the interest he attaches to the design of the CQF emitters through his behavior and attitude. He must show that the information released will produce real added value to give rise to competitiveness factors. He must also show that he is ready to overcome the challenges which have emerged. It is essential that the motivation of the processor's human resources is maintained at a high level throughout the process.

At the end of the process, the project manager is to carry out a report on the initial relevance of the performance of these emitters, the level of consensus in the choices made and the ability of team members to then work together in complying with the production of these specifications.

6.7. Conclusion

CQF emitters, like the CQF that they release, constitute the core of the innovation project's communication system with various stakeholders within the driver of socioeconomic exchanges, in order to "sell" the intrinsic quality produced by the innovation process so as to maximize the project's chances of success.

Mastering these communication methods is therefore essential. However, the fact that it is a complex procedure and that a considerable amount of time must be devoted to it without it creating any intrinsic value to the innovation process means that project managers and other team members tend to neglect it. It is difficult to create and to then communicate about its creation. Nevertheless, it is through this type of communication that the socioeconomic success of the project can be ensured.

This specific type of communication to "sell" the intrinsic quality of an innovation process is becoming more and more difficult to carry out. Projects are faced with an ever-growing number of competitors vying for the same stakeholders. The stakeholders demand more and more information before agreeing to collaborate on a project. Parties associated with the driver of socioeconomic exchanges have an increasingly significant influence on stakeholders' competitive quality judgments.

It can sometimes be useful to use the services of communication specialists to help master this type of quality evaluation. However, they cannot replace the project manager or the quality department working alongside, in implementing their own methodologies for this purpose.

7

Qualification of Emitters

7.1. Introduction

The specifications for the information which forms the CQF for various classes of stakeholders in an innovation process evolve as the process advances. The emitters of this information must adapt to these evolutions without modifying their basic initial configuration. Since these specific processors will be used several times, with the same basic configuration, it can be useful to test whether this basic configuration is able to release information in compliance with the specifications imposed, before they are used for the first time. This stage of the competitive quality process (CQP) is known as the "qualification" of the competitive quality fuel (CQF) emitters.

The necessary flexibility of an innovation process means that it must be prepared for an evolution in the configuration of the CQF emitters without modifying their structural base. Only the following stages in the CQP, which consist of checking the capability of the CQF emitters and stakeholders' perceived quality, can cause these basic structures to be modified.

The goal of this chapter, organized in the same way as the previous ones, is to provide guidelines for carrying out this qualification stage.

7.2. The qualification of competitive quality fuel emitters

This qualification is a preventative process intended to ensure these emitters are able to communicate information in compliance with the imposed specifications, when they are used for the first time.

These emitters must obtain, shape and communicate information compliant with the given specifications. It can be described as follows:

– obtaining the information means extracting certain data in its raw state from its source (primary information);

– shaping the information means creating a presentation of the extracted data which complies with the specifications;

– communicating the information means placing this presentation within reach of stakeholders, making it accessible while still remaining compliant with the specifications.

Qualifying the information means testing, as soon as it is designed, whether a CQF emitter is able to:

– extract the necessary primary information from their source;

– transform this information to ensure that it complies with the CQF specifications;

– make this presentation attractive for stakeholders' quality comburant.

7.3. Method for qualifying competitive quality emitters

In order to offer guidelines showing a method for qualifying a CQF emitter, we will start with an example.

We shall return to the example of creating a production system for made-to-measure artificial hip joints adapted to each patient's physiology, where the stakeholder in question is the *Haute Autorité de Santé* (HAS), or French National Authority for Health. One of the pieces of information demanded by the HAS is a report by a registered body certifying that each made-to-measure joint is under stringent enough checks to ensure that every joint will meet the demands for CE marking.

This report is the result of an information emitter who in this case is the auditor from the registered body.

The audit is planned for the end of the innovation process when the production system for the made-to-measure artificial hip joints has been qualified.

In reality, the project manager will have carried out regular internal planned audits at various stages during the evolution of the innovation's configuration to evaluate the situation of the information production to be used during the audit by the registered body. It is the responsibility of the company's regulatory affairs department to carry out this series of fragmentary internal audits.

The design of this sequence of emitters is guided by European regulations, but also by a specific auditing methodology systematically applied in the same way for each sequence.

So as not to be both judge and jury, the project manager delegates the design and the activation of this emitter to a party outside the project team, or even outside the company altogether.

The qualification of these emitters is carried out on several levels.

After each internal audit, the department of regulatory affairs must self-evaluate the performance of its own work by identifying, above all, the difficulties encountered in:

– accessing certain factual information representative of the situation under scrutiny;

– shaping the collected data so as to stimulate the project team's quality perceptions;

– preventing confidential information being released while satisfying the regulatory demands of CE marking.

Subsequently, the presentation of internal audit reports during some project reviews must allow the project manager to complete the qualification of internal audits.

The qualification of the audit by the registered body is guaranteed by the accreditation of this body.

It is too frequently observed that many information emitters within a company are neither qualified nor subject to controls, even though the play a fundamental role in the decision-making process.

Generally speaking, the qualification of a CQF emitter consists of the following stages:

– prompting the activation of the emitter by using the most representative source of information possible, from which raw data will be drawn;

– observing the methods of extraction of primary information necessary for the communication of information which complies with the CQF specifications imposed;

– identifying inefficiency risks in obtaining this information and estimating their criticality;

– formulating any corrective actions necessary to reduce the criticality of these risks to acceptable levels;

– validating the implementation of these corrective actions;

– producing a qualification report;

– reproducing the same process for the stages involving the transformation of the extracted data to ensure that they comply with CQF specifications, and for the stages where this information is made available to stakeholders in order to stimulate their CQF as the competitive marketing quality advocates.

The concept of risk criticality, as mentioned above, can be defined in the following way:

– the criticality of a risk is estimated using the following formula:

$$C(r) = [\Sigma_i (P (e_i/r) G (e_i)]$$

– where:

- $C(r)$ is the criticality of the risk r;

- $P (e_i/r)$ is the conditional probability of witnessing an effect e_i when risk r occurs;

- $G (e_i)$ is an estimate of the magnitude of the effect e_i.

These actions are often guided by procedures which must be systematically complied with in order to be able to detect a lack of efficiency in the innovation process as soon as possible.

It is not always necessary to carry out a qualification of a CQF emitter. The production of certain emitters is perfectly controlled because they are reproduced, in a robust and identical way, in every innovation project in the company. However, this should be approached with caution since it often arises from a lack of will to plan such qualifications, which may be deemed time-consuming.

There is a strong tendency to only carry out these qualifications when there is a high danger of not being able to produce CQF which complies with its definition.

As the example shows, these qualification stages must be planned at precise intervals during the innovation process. This planning must be carried out systematically after the emitters are designed and before they are implemented for real.

7.4. The performance of this qualification

A qualification is a way of filtering out the risks of not complying with the objectives established following the design of the processor responsible for producing the desired results.

The measurement of performance is done in two stages:

– when the qualification report is submitted;

– after the production of CQF by each qualified emitter.

During the qualification process, it is necessary to highlight the risks of non-conformity which are at an acceptable level of criticality at the qualification stage, but which must be controlled during the actual production of the processor to prevent this criticality rising to unacceptable levels.

The first partial evaluation of qualification performance is carried out when the qualification report is submitted. It may be noted at this stage that the various stages of the qualification process have not been correctly followed. It is also possible that some team members are showing a lack of confidence in the ability of the emitter to communicate credible information which complies with the CQF specifications. When such a drop in confidence is perceived, the emitters in question must be audited and re-qualified without delay.

The second part of the evaluation is carried out by observing non-conformities in the production of CQF and by assigning the causes to a shortfall in the identification of critical risks of non-conformity during the qualification of the CQF emitter.

7.5. Managing the qualification of competitive quality fuel emitters

The qualification of CQF emitters is a process which is led by the project manager who is solely responsible for the efficiency of the quality process.

However, due to the methodological specificities and to prevent the managerial department being both judge and jury, it is necessary to entrust the various operational stages to a member of the company's quality department working on the project.

The relations which must be implemented between the various members participating in the qualification process can be structured as per the following model.

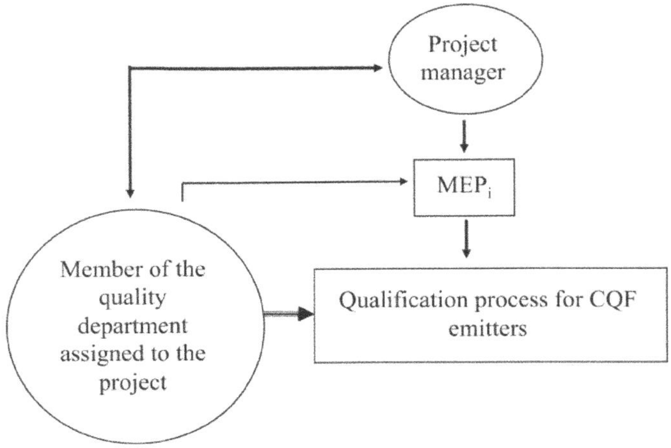

Figure 7.1. *Managerial organization of a processor for qualifying competitive quality fuel emitters*

COMMENT ON FIGURE 7.1.–

– MDA_i denotes the members of the management within the company responsible for relations with a specific class of stakeholders (human resources management, financial service management, marketing management, purchasing management, administrative management, legal department, etc.).

– MEP_i denotes the members of project team number (i) who contribute to the added value of the innovation process by supplying a particular type of skills related to an area of expertise.

These members contribute to the design of the CQF emitters.

Relations between project manager and other members of the processor are, as before, managerial relations by delegation. The various members who participate alongside the project manager are seconded to the project. They belong to other "managerial areas" and made available for the sake of the project. The project manager gives orders to these members.

He delegates the operational responsibility for carrying out the process to a member of the quality department made available to the project, and provides him with the resources to fulfill this role. The other members carry out the orders of the project manager under the operational responsibility of the member of the quality department.

This member will adopt "client–provider"-type relations with the MDA_i and MEPi, under the managerial responsibility of the project manager. The member of the quality department adopts the role of the client. The compensation for these providers is the submission from the quality department of the qualification of the emitters which they help to design.

The initial meetings are generally dedicated to ensuring that the processor which carries out the qualification is able to fulfill its task. The representative from the quality department ensures that the members of this processor share the opportunity and feasibility of such a process. He also validates the specific scenario regarding the process which he has designed or which complies with pre-established procedures. He distributes roles among the various team members.

The member of the quality department is accountable for the results of the qualification process during the project reviews.

The project manager polices any conflicts of interest which may arise between those involved in this process.

He must always demonstrate the interest he attaches to the qualification of these CQF emitters.

7.6. Conclusion

We devoted this entire chapter to the process of qualifying CQF emitters because they play a key role, which is often underappreciated, in the effectiveness of quality processes integrated into the innovation process.

A business plan which does not emphasize the economic potential of an innovation process can, for instance, lose part of the financial resources essential for its success.

Poor internal communication can demotivate members of a project team and others.

This stage of the CQP, like many others, is often considered by members of the project team as being pointlessly time-consuming. They are more attracted to the technical innovation performance by its effects on quality perceptions of stakeholders.

The qualification of these vital information emitters is a way of bringing together various team members through collective action, which allows them to take stock of their current practices.

8

Validation of Marketing Quality

8.1. Introduction

After ensuring that the innovation process is able to release competitive quality fuel (CQF) in compliance with the predefined specifications, it is necessary to verify that the various scenarios envisaged by competitive marketing quality are able to be carried out in order to limit the risks of failure.

This aspect of the competitive quality process (CQP) is the equivalent of a movie premiere.

For the first time, the hard facts of the exchanges are shown to the innovation project's stakeholders.

For relations with each class of stakeholders within the innovation process, this corresponds to the first full-scale cycle of complete exchanges.

This also requires the project manager to initiate a first relational sequence with a sample of stakeholders which starts off the following process:

– release of CQF for the products or services offered to a given class of stakeholders;

– expression on competitive quality judgments of these products or services by the stakeholders;

– contributions of stakeholders to the success of the innovation process;

– contributions of stakeholders to the company's reputation regarding its ability to innovate.

This sequence can be used to test the performance evaluation system of the various stages of the CQP.

8.2. Validation of the competitive quality marketing of an innovation process

This validation is a full-scale check of the feasibility of the competitive marketing quality scenarios of the previous four processes and an additional attempt to optimize their efficiency.

It is carried out during the first exploratory exchanges with stakeholders.

The four processes are overseen by observers who participated in designing the scenarios.

The observers evaluate the initial performance of these processes when they apply the competitive marketing quality scenarios.

An end of observation report is examined during some project reviews. It can lead to decisions being made to undertake modifications in the scenarios in order to reduce the criticality of the risks of non-compliance with the scenarios or to grasp opportunities for progress noted during these observations.

These observations also help to lay the methodological foundations for the subsequent controls necessary for the various processes.

This validation stage is also part of the preventative actions in the CQP. It is necessary because innovation processes are “one-shot” groundbreaking procedures.

8.3. Methods for validating competitive marketing quality in an innovation process

The validation process must first be prepared by:

– choosing a stakeholder who will fulfill the role of “guinea pig”. This choice must be taken limiting the risk of losing an important partner in the innovation process, but ensuring that the test is sufficiently representative of the situations which will be encountered later on;

– nominating observers based on their ability to spot the critical points in the four processes. These observers must be able to observe the situation from various angles. They must be just as perceptive at identifying and creating a hierarchy of risks, as spotting new opportunities emerge. They must be familiar with the scenarios envisaged for the processes to be observed, and the specifications attributed to the CQF and their emitters;

– directing observations with guidelines established during the previous stages of the CQP and by appropriate methodological tools.

Subsequently, it is necessary to plan this validation at the correct time in the innovation process. If it is planned too early, it risks not providing observers with the correct guidelines. If it is too late, burdensome and demotivating corrective action may have to be taken.

It is essential to prevent bias being introduced during these observations by insisting the team members involved in the process to adopt an attitude that they could not reproduce later on.

After each of the four processes is activated, a summary of observations must be produced to give an overview of how the scenarios of these processes are followed. This summary must only include the key facts without creating a critical analysis. After which, the observer group must collectively carry out a critical analysis linking the facts featured in the overview to probable effects and possible causes. A hierarchy of FMECA-type methods then helps to rank the contribution of facts to the non-compliance with provisions of the initial CQT.

This ranking guides the decisions of the project manager on whether to undertake a process to improve the performance of various previous stages in the CQP.

8.4. Planning the validation of competitive marketing quality

In the previous section, it was shown that the planning of this stage of the CQP has great importance.

We propose the following guidelines for this process:

Stakeholders in the CQT	Stages of the innovation process when it is carried out.
Future clients of products or services which will incorporate the innovation	Conducting prospective studies on possible socioeconomic uses for the innovation, and its impacts on the company and its position.
Potential investors in the innovation and other stakeholders contributing to the project	Conducting studies on the nature and feasibility of exchanges with future stakeholders in the project necessary for it to be fully realized.
Future contributors to the business unit who will produce and distribute the products or services incorporating the innovation	Conducting tests on the drafts of future products or services, at various steps in their production, on future clients and on all future stakeholders in the production system for the products and services.

Box 8.1. *Stages of the innovation process during which the competitive marketing quality is validated*

8.5. Managing competitive marketing quality validation

This validation is also a process which is led by the project manager.

For the same reasons as previously mentioned, the running of the various operational stages must be entrusted to a member of the company's quality department which is following the project's progress.

The relations which must take place between various participants in this process are modeled in the following figure.

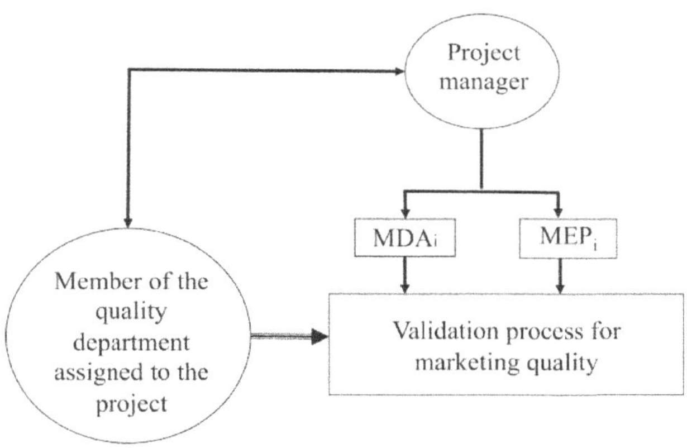

Figure 8.1. *Managerial organization of a processor for validating the competitive marketing quality*

COMMENT ON FIGURE 8.1.–

– MDA_i denotes the members of the management within the company responsible for relations with a specific class of stakeholders (human resources management, financial service management, marketing management, purchasing management, administrative management, legal department, etc.).

– MEP_i denotes the members of project team number (i) who contribute to the added value of the innovation process by supplying a particular type of skills related to an area of expertise.

These members contributed to designing the competitive marketing quality.

Relations between project manager and other members of the processor are, as stated before, managerial relations by delegation. The various members who participate alongside the project manager are seconded to the project. They belong to other "managerial areas" and made available for the sake of the project. The project manager gives orders to these members.

He delegates the operational responsibility for carrying out the process to a member of the quality department made available to the project, and provides him with the resources to fulfill this role. The other members carry out the orders of the project manager under the operational responsibility of the member of the quality department.

This member will adopt "client–provider"-type relations with the MDA_i and MEP_i, under the managerial responsibility of the project manager. The member of the quality department adopts the role of the client. The compensation for these providers is the submission from the quality department of the qualification of the emitters which they helped to design.

The initial meetings are generally dedicated to ensuring that the processor is able to validate the marketing quality. The representative from the quality department ensures that the members of this processor share the opportunity and feasibility of such a process. He also validates the specific scenario regarding the process which he has designed, or which complies with pre-established procedures. He distributes roles among the various team members.

The member of the quality department reports on the results of the validation process during the project reviews.

The project manager polices any conflicts of interest which may arise between those involved in this process.

He must always demonstrate the interest he attaches to the validation of marketing quality.

8.6. Conclusion

This part of the CQP is not time-consuming because it takes place at the same time as the initial exchanges with each class of stakeholders. However, it requires the introduction of a specific methodological process of observation and use of information. This information is released during the four processes of the production and use of stakeholders' perceived quality which relies on the competence of the quality department.

It also requires members of the project team to listen to the beneficiaries of the products or services that they are offering to stakeholders in the innovation process. This attitude may seem unnatural. When an original idea is developed, energy is concentrated on creative research which is perceived as being at the heart of value creation. Reactions of future stakeholders are often seen as potential disruptions. The member of the quality department who is thrown into the midst of the creators is often seen as someone who spots problems at bad moments when he attracts the attention of project team members to the need to react quickly to stakeholders' initial quality perceptions.

9

Supervision of Emitters

9.1. Introduction

Some stages that are addressed seem basic components of a quality process integrated into a mass production process, but it is much more complex to implement them into an innovation process.

For each class of stakeholders, the supervision of competitive quality fuel (CQF) emitters, during the release of information of which the emitters retain an acceptable capacity level, consists of ensuring compliance with the specifications demanded of this information and with the marketing quality scenarios, and then eliminating the elements of non-compliance as soon as possible. This supervision also consists of seizing opportunities to improve the performance of these emitters.

This is not a case of identical reproduction because the information comes from a "one-shot" process. Each emitter has been designed once and for all. They release information in sequences and are qualified during a first experimental sequence of information release. It is then a question of ensuring: that the same capacities are retained from one production sequence to the next; that areas of non-compliance to predefined specifications of information released are corrected as soon as possible, and if possible, before they have an effect on stakeholders' competitive quality perceptions; and that they remain flexible enough to take advantage of significant improvements.

9.2. The objective of supervision

The objective of this supervision is to prevent a deterioration of stakeholders' competitive quality perceptions by CQF information which does not comply with the predefined and validated specifications, or which does not comply with the marketing quality scenarios. It is also an opportunity to take advantage of the observations carried out by this supervision to improve the functioning of these emitters.

The goal is, therefore, to avoid stakeholders' quality judgments which would stop them from collaborating as planned in the success of the innovation process because the CQF information does not comply with their specifications. This aim is also to improve the efficiency of these emitters when possible. This process cannot completely guarantee that no uncompetitive quality judgments will arise despite the validation of the marketing quality process for the production/use of stakeholders' perceived quality carried out previously. However, it helps to eliminate the majority of rational causes not due to evolutions in stakeholders' quality comburants which are hard to predict.

This supervision consists of three lines of action:

– the first is preventative: it involves reducing the risk of releasing information that does not comply with specifications to acceptable levels;

– the second is corrective: this involves correcting areas of non-compliance before they are perceived by stakeholders, or compensating, as soon as possible, for the effects on their competitive quality perceptions caused before uncompetitive quality judgments are expressed;

– the third is progressive: it capitalizes on opportunities to improve the efficiency of these emitters in real time using observations carried out.

9.3. Methods of supervision

This is a traditional supervision method.

For the preventative stages, it is necessary to verify, before each information sequence of a CQF emitter, that its configuration complies with its design specifications, and that the source of primary information which it must obtain and transform is easily accessible. When risks of releasing CQF that does not comply with these specifications are identified, and the criticality level of these risks is too high, preventive corrective actions must be taken.

It should be recalled that the "configuration" of a system is made up of the structure of its resources, which may be human, material, information-based, organizational and financial, along with its functional potential, which is the collection of its functionalities and the time-based organization of its actions (process).

For the corrective stages:

– the information released by the emitter is observed;

– it is compared with the CQF specifications which it must release;

– decisions are taken on how to correct the deviations observed;

– these corrections are carried out;

– these corrective actions are recorded in order to build knowledge and expertise.

For actions to bring about improvements, it is necessary to pursue all areas of improvement as quickly as possible provided that they do not destabilize the methods of CQF information release.

The above steps are intentionally impersonal because the key difficulty of this supervision stage lies in specifying who takes which role.

We recommend a systematic "self-supervision" at this stage of the competitive quality process (CQP), where the emitter is responsible for the supervision.

The corrections are made to either:

– the CQF itself, making it comply with the specifications before it has an effect on stakeholders' quality comburant;

– the causes that may lie in the way the emitter was activated, or in its configuration;

– or the compensation for the effects caused to stakeholders' competitive quality perceptions.

9.4. Planning the supervision of competitive quality fuel emitters

As previously mentioned, we propose the following guidelines for the planning process.

Category of stakeholders in the CQT	Stages in the innovation process where this is addressed
Future clients of the products and services incorporating the innovation	– Conducting initial feasibility tests on the future products or services – Characterizing future uses for products or services that would incorporate the innovation – Designing drafts of the production and distribution systems for the products and services – Conducting studies on the ability of the company to design production and distribution systems for the products and services that will incorporate the innovation by taking into account the configurations of the various definitive drafts of the innovation – Conducting tests on the drafts of future products or services, at various steps in their production, on future clients and on all future stakeholders in the production system for the products and services.
Potential investors in the innovation and other stakeholders contributing to the project	– Conducting studies on the nature and feasibility of exchanges with future stakeholders in the project necessary for it to be fully realized – Building a processor responsible for carrying out the innovation process and an external communication network – Characterizing future uses for products or services that would incorporate the innovation – Conducting prospective studies on possible socioeconomic uses for the innovation, and its impacts on the company and its position – Conducting initial feasibility tests on the future products or services

	– Conducting tests on the drafts of future products or services, at various steps in their production, on future clients and on all future stakeholders in the production system for the products and services
Stakeholders in the BU's strategic area of activity that will produce and distribute the products or services incorporating the innovation	– Conducting studies on the ability of the company to design production and distribution systems for the products and services that will incorporate the innovation by taking into account the configurations of the various definitive drafts of the innovation – Designing drafts of the production and distribution systems for the products and services – Conducting tests on the drafts of future products or services, at various steps in their production, on future clients and on all future stakeholders in the production system for the products and services

Box 9.1. *Stages in the innovation process in which the competitive quality fuel emitters are supervised*

9.5. The performance of the supervision of CQF emitters

The performance of this supervision has several stages, measured in a traditional way by evaluating:

– the number of areas of non-compliance observed, giving an indication of the performance of the preventative action taken;

– the number of impacts on stakeholders' immediate quality perceptions of these areas of non-compliance, giving an indication of the performance of corrective action intended to eliminate non-compliance before they affect quality perceptions;

– the number of sustainable impacts of non-compliance on stakeholders' quality judgments, giving an indication of the performance of the compensation intended to override the negative quality perceptions;

– the number and relevance of the efficiency improvements to the emitters as a result of this supervision.

9.6. Managing the supervision of competitive quality fuel emitters

This supervision process is also led by the project manager.

Members of the project team must carry out self-supervisions.

The member of the company's quality department working on the project simply assists the team members in terms of methodology.

The relations that must take place between various team members who participate in this process are shown in Figure 9.1.

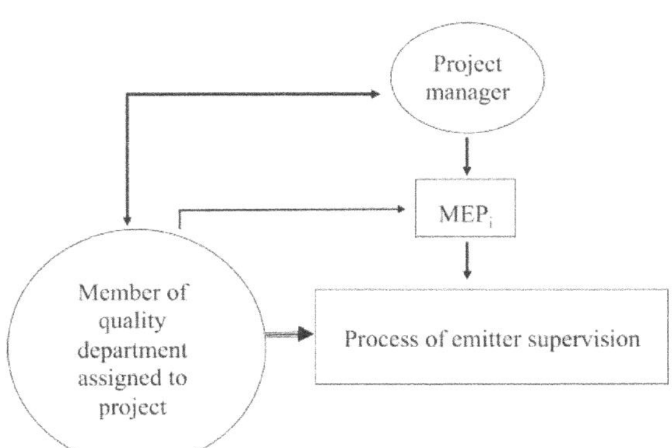

Figure 9.1. *Managerial organization of a processor for supervising competitive quality fuel emitters*

COMMENT ON FIGURE 9.1.– MEP$_i$ denotes the member of project team number (i) who contributes to the added value of the innovation process by supplying a particular type of skills related to an area of expertise.

Relations between project manager and other members of the processor are, as before, managerial relations by delegation. The various members who participate alongside the project manager are made disposition to the project. They belong to other "managerial areas" and made available for the sake of the project. The project manager gives orders to these members.

They delegate to each one the operational responsibility for the self-supervision of the CQF emitters which it is their role to activate. They request the member of the quality department to be available to assist in the methodological aspect of this process.

This member will adopt "client-provider"-type relations with the MEPi, under the managerial responsibility of the project manager. The member of the quality department adopts the role of the provider. The compensation is sharing with the rest of the project team members the efficiency of the CQF emitters.

The initial meetings are generally devoted to ensure that the project team members are ready to carry out the self-supervisions by asking for advice from the quality department representative.

9.7. Conclusion

This aspect of the CQP is difficult to carry out because the CQF specifications and CQF emitters are often poorly defined. This supervision process is more robust when these specifications address the demands formulated by the stakeholders who are the beneficiaries of the products or services produced by the project.

The "one-shot" sequential nature of the method for producing the quality process, might suggest that this stage is of little use. This would be forgetting that the dissemination of information that does not comply with the CQF can have disastrous consequences on the expected benefits from exchanges with stakeholders. These effects are amplified when a project involves a large number of stakeholders, and a non-compliance in a CQF can have a domino effect because of a series of devastating situations. This situation has been seen in innovation process assisted by computer network. Furthermore, the significant risk that the stakeholder is taking in an innovation process increases their sensitivity to their quality comburants. The slightest deviation from certain CQF specifications can lead to a rapid loss of confidence from the stakeholders in the ability of the project to deliver on its commitments. Therefore, in order to "gamble" on an innovation project, it is necessary above all to be able to have confidence in the project promoter. Robust, simple and personalized methodological support from the quality department representative is essential to give confidence to all project team members in their ability to usefully and effectively implement this self-supervision approach.

10

Monitoring Perceived Quality

10.1. Introduction

Monitoring the impact of competitive quality fuel (CQF) in the innovation process on various stakeholders' competitive quality perceptions is often limited to taking account of their emergence. The sensitivity of stakeholders' quality comburants, as highlighted in the previous chapter, is more likely to produce such notifications than in other situations. Nevertheless, they often represent only the tip of the iceberg in terms of their uncompetitive quality perceptions. The signals given during this process often mask a deep-rooted dissatisfaction which can considerably harm the company's ability to reap the benefits expected.

Even though this process goes against the creators' natural instincts who often think that the disruption caused by the innovation prevents stakeholders from truly appreciating their work, it is essential to obtain as much information as possible on CQF impacts on the competitive quality perceptions of various stakeholders. The more significant the changes, the more stakeholders seem to take the high criticality risks, putting their confidence in the benefits declared by the members of the innovation project. This stage of the competitive quality process is, therefore, particularly important and sensitive to carry out.

In summary, it can be said that:

> *The members of the innovation project are not naturally "client oriented", and the stakeholders in the project are not initially inclined to put their trust in the declaration of benefits formulated by the project in order to encourage their involvement stakeholders.*

10.2. The objective of this monitoring

The objective of this monitoring is therefore to understand, as quickly as possible, the positive or negative effects of the CQF released on the stakeholders' competitive quality perceptions, guided by the objectives defined in the competitive quality tactics, and to anticipate the likely impacts of the expected benefits from the project due to its involvement.

This monitoring also includes the actions which help modify the directions of the process for communicating the CQF in order to take advantage of new opportunities to increase the competitiveness of the innovation through stakeholders' quality judgments, or to correct the deviations from the objectives in the competitive quality tactics.

The innovation is a projection into an uncertain future. The uncertainty reduces gradually as the information is released regarding how the exchanges with stakeholders will take place. In particular, stakeholders' competitive quality perceptions evolve as the CQF emitters release their information. As long as they are releasing information, it is possible to influence these perceptions in order to promote exchanges with stakeholders which would allow the company to obtain the resources it expects.

10.3. Methods of monitoring perceived quality

This monitoring is, as discussed in the previous chapter, known as a "control loop". It is a particular process which consists of the following stages:

– observing the signals which show stakeholders' competitive quality perceptions;

– interpreting this information with regard to the contracts negotiated and the competitive quality tactics adopted;

– deciding any direction changes which need to be made in the process of releasing CQF;

– modifying, if necessary, CQF specification and the configurations of their emitters in order to carry out these direction changes;

– requalifying the emitters of the modified CQF;

– characterizing the impacts obtained on stakeholders' quality perceptions.

What is new in this process is the collection of information on stakeholders' competitive quality perceptions. How can this collection be carried out?

First, it is clear that the members of the department, within the company who are made available for the project, responsible for relations with the class of stakeholder in question, must carry out this task in as natural way as possible. They are used to observing stakeholders' attitudes and behavior. In general, they know how to identify the signals which draw their attention toward the emergence of these perceptions. They can then supplement this information by engaging in a targeted dialogue with the stakeholder. This expertise can be amplified by training based on simulations of situations, or the use of appropriate methodologies for observing and questioning.

Another important aspect of this type of monitoring is the way in which the collected information is interpreted. Project members are so focused on their lines of action in the innovation that they struggle to put themselves in the stakeholders' shoes in order to identify and understand the causes of these perceptions. This change is very difficult and must be led by those project team members who are in contact with stakeholders, assisted by the representative of the quality department involved in the project.

The different viewpoints of these team members can be a source of conflict, which must be quickly resolved by the project manager.

Common sense and expertise play a key role in this stage of monitoring.

10.4. Planning the monitoring of perceived quality

The planning of the monitoring of perceived quality during the various stages of the innovation process can be summarized as per Box 10.1.

Category of stakeholders in the CQT	**Stages in the innovation process when it is carried out**
Future clients of the products and services incorporating the innovation	– Conducting prospective studies on possible socioeconomic uses for the innovation, and its impacts on the company and its position – Conducting tests on the drafts of future products or services, at various steps in their production, on future clients and on all future stakeholders in the production system for the products and services – Conducting performance tests on future products or services
Potential investors in the innovation and other stakeholders who contribute to the project	– Briefly characterizing products or services which could benefit from the innovation – Identifying possible clients for these future products or services – Conducting prospective studies on possible socioeconomic uses for the innovation, and its impacts on the company and its position – Conducting studies on the nature and feasibility of exchanges with future stakeholders in the project necessary for it to be fully realized – Building a processor responsible for carrying out the innovation process and an external communication network – Creating a rationale to aid decision-making, examining the probable socioeconomic impacts of implementing the innovation – Conducting performance tests on future products or services – Conducting tests on the drafts of future products or services, at various steps in their production, on future clients and on all future stakeholders in the production system for the products and services

	– Conducting studies on the ability of the company to design production and distribution systems for the products and services which will incorporate the innovation by taking into account the configurations of the various definitive drafts of the innovation
Future contributors to the business unit who will produce and distribute the products or services incorporating the innovation	– Conducting prospective studies on possible socioeconomic uses for the innovation – Characterizing future uses for products or services which would incorporate the innovation – Conducting performance tests on future products or services – Conducting studies on the ability of the company to design production and distribution systems for the products and services which will incorporate the innovation by taking into account the configurations of the various definitive drafts of the innovation – Conducting tests on the drafts of future products or services, at various steps in their production, on future clients and on all future stakeholders in the production system for the products and services – Designing drafts of the production and distribution systems for the products and services

Box 10.1. *Stages of the innovation process where perceived quality is monitored*

10.5. Managing the monitoring of perceived quality

This monitoring process is also led by the project manager.

Members of the project team in contact with the stakeholders must be given the role of carrying out the various operational stages.

The relations which must take place between the various participants in this monitoring are shown in Figure 10.1.

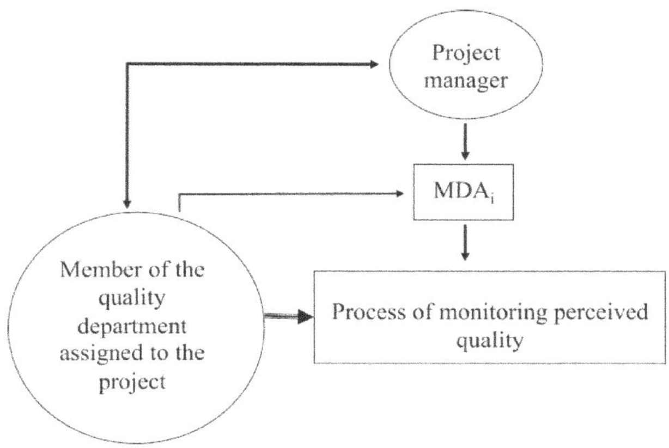

Figure 10.1. *Managerial organization of a processor for monitoring perceived quality*

COMMENT ON FIGURE 10.1.– MDA$_i$ denotes the member of the management within the company, responsible for relations with a specific class of stakeholders (human resources management, financial service management, marketing management, purchasing management, administrative management, legal department, etc.)

These team members contributed to the design of the competitive marketing quality.

Relations between project manager and other members of the processor are, as stated before, managerial relations by delegation. The various members who participate alongside the project manager are made disposition to the project. They belong to other "managerial areas" and are made available for the sake of the project. The project manager gives orders to these members.

They delegate the operational responsibility for carrying out the process to a member of the quality department made available to the project, and provide them with the resources to fulfill this role. The other members carry out the orders of the project manager under the operational responsibility of the member of the quality department.

This member will adopt "client–provider"-type relations with the MDA_i, under the managerial responsibility of the project manager. The member of the quality department adopts the role of the provider. They provide the MDA_i with the methodological support to carry out this monitoring. The compensation which they receive from them is the quality perceived by the project manager of this monitoring. The representative from the quality department ensures that the members of this processor share the opportunity, and feasibility of such a process. They also validate the specific scenario regarding the process which they have designed. They distribute roles among the various team members.

10.6. Conclusion

The monitoring of stakeholders' perceived quality in an innovation process is shared out among those project team members who are most naturally in contact with the stakeholders. This monitoring is one of the guides of the innovation process. The CQF emitters release information in a sequential manner. After each sequence, stakeholders' competitive quality perceptions evolve. By monitoring these evolutions, the project manager can adapt the direction of this information in order to strengthen the chances of eventually obtaining competitive quality perceptions which comply with the objectives of the competitive quality tactics. This monitoring can sometimes lead to a modification of the tactics themselves.

It is frequently observed that this monitoring is difficult to carry out and interpret because the emitters only release incomplete information so as not to reveal information which could feed the competition. This incomplete information can frustrate stakeholders and artificially influence their competitive quality perceptions. It is, therefore, necessary to distinguish non-quality perceptions due to this frustration from those which are caused by CQF specifications reflecting certain aspects of the innovation process to be modified.

In an innovation process, exchanges between stakeholders are often coproducts of products or services, and are therefore in the field of "service process". This particular situation, often observed in "one-shot" processes, enables real-time reaction when non-competitive quality perceptions are spotted. This particular flexibility, linked to the "task force" organization of the project teams, gives further weight to the process of monitoring perceived quality.

11

Ensuring Benefits

11.1. Introduction

Irrespective of the exchanges that take place with stakeholders, it is important to ensure, as quickly as possible, that the compensation which they agree to provide to the project or the company complies with the specifications initially negotiated with these stakeholders.

A stakeholder can be very satisfied with competitive quality fuel (CQF) information, and yet does not provide the compensation expected by the project or the company. On the other hand, a stakeholder can have doubts about the credibility of information in the CQF targeted at it, without this being spotted by those responsible for monitoring their perceived quality. These doubts can be the source of compensation which does not comply with the project manager's expectations, and leads to the failure of the project.

It should be recalled that an innovation process can be destabilized, and it does not achieve its objective as soon as a stakeholder fails to play the role previously negotiated.

It is, therefore, necessary to quickly be able to spot signals which indicate a critical risk of not obtaining the benefits expected from the stakeholders.

11.2. The objective of ensuring benefits

The objective of this stage in the competitive quality process integrated into an innovation process is, therefore, to estimate as soon as possible in exchanges with the stakeholders the chances of obtaining (or not obtaining) the benefits which have been agreed upon, sometimes in the form of a contract.

During the innovation process, several factors can modify the relations initiated with a stakeholder. A relation is initiated on the basis of a presentation of a project, which is a projection into an uncertain future. The initial CQF information released, which helps to structure the relation in the form of a contract, gives rise to an image of this projected reality among the stakeholders. Each one has their own vision of this projected reality, even though the contract aims to create a common platform to standardize these visions.

The CQF released during the innovation process and the information obtained by the project team members on changes in the stakeholders' competitive quality perceptions cause these visions to evolve, and can lead to a modification in their estimates of the chances to obtain the benefits expected. This can even lead to the initial contract being reconsidered.

The quicker this risk is identified in the process, the easier it is to act on it to ensure that the level of criticality can be reduced to acceptable levels or to reconsider the rules in the interaction with the stakeholder to limit any negative effects.

11.3. Methods for ensuring benefits

This process of ensuring benefits is also, as seen previously, a "control loop". It consists of the following stages:

– obtaining information which describes signals of emerging risks that stakeholders will not provide the expected compensation;

– estimating the criticality of this risk;

– if necessary, carrying out a diagnosis of causes to better understand the situation;

– defining the corrective action necessary to reduce the criticality of the risk to acceptable levels;

– carrying out this corrective action;

– evaluating the impact on the criticality of the risk.

The new aspect on this process is the step of obtaining information on the attitude or behavior of the stakeholders, which indicates the emergence of a risk, with an unacceptable criticality level, of not receiving the benefits expected from them.

How is this information obtained?

As stated before, this information is obtained by those members of the directorate within the company who are responsible for relations with the class of stakeholders in question (directorates of human resources, financial service, marketing, purchasing, administration, legal, etc.).

These are members who are most frequently in contact with the stakeholders who must carry out this task, in as natural a way as possible. They cannot be members of the project team, but they are temporarily placed under the responsibility of the project manager at their request. They are accustomed to observing stakeholders' attitudes and behavior. In general, they know how to identify the signals which attract their attention to a risk. They supplement this information by forging a targeted dialogue with the stakeholders.

It is equally important in this type of process to be able to interpret the information collected.

Usually, in this particular situation, when the interpretation of the information collected is obvious, an apparent risk can be linked to a single effect which is perceived as being the most serious.

In more complex cases, it is necessary to introduce this more technical method in order to estimate the criticality of a risk.

The representative of the quality department within the project team is responsible for carrying out this type of task, and to propose any tools which may be needed to facilitate this work.

When a diagnosis of causes is necessary, it is carried out by the representative of the quality department.

This representative is also responsible for carrying out the process of corrective actions.

11.4. Planning methods for ensuring benefits

The various stages of the process for ensuring benefits for the project or company are planned according to the ways mentioned in Box 11.1, within the innovation process.

Category of stakeholders in the CQT	**Stages in the innovation process when it is carried out**
Future clients of the products and services incorporating the innovation	– Conducting prospective studies on possible socioeconomic uses for the innovation, and its impacts on the company and its position – Conducting tests on the drafts of future products or services, at various steps in their production, on future clients and on all future stakeholders in the production system for the products and services – Conducting performance tests on future products or services
Potential investors in the innovation and other stakeholders who contribute to the project	– Briefly characterizing products or services which could benefit from the innovation – Conducting studies on the nature and feasibility of exchanges with future stakeholders in the project necessary for it to be fully realized – Identifying future stakeholders in the innovation project – Building a processor responsible for carrying out the innovation process and an external communication network – Conducting performance tests on future products or services – Conducting studies on the ability of the company to design production and distribution systems for the products and services which will incorporate the innovation by taking into account the configurations of the various definitive drafts of the innovation
Future contributors to the business unit who will produce and distribute the products or services incorporating the innovation	– Conducting tests on the drafts of future products or services, at various steps in their production, on future clients and on all future stakeholders in the production system for the products and services

Box 11.1. *Stages of an innovation process in which benefits are ensured for the company*

11.5. The management of ensuring benefits

This stage of the competitive quality process is also led by the project manager. They delegate the operational running of the process to the representative of the quality department.

This representative leads the group made up of members of the directorates within the company, responsible for relations with the classes of stakeholders in question.

They are the team members who observe the attitudes and behavior of the stakeholders in order to spot the signals, as quickly as possible, of risks of not obtaining the benefits expected of them.

As soon as these signals are detected, they alert the quality department representative who helps them to carry out an initial rapid estimate of the criticality of these risks.

If necessary, the representative then alerts the project manager in order to decide how to proceed in this situation.

The representative of the quality department leading the process implements the decisions of the project manager, involving those team members who made the initial alerts.

Finally, They help them to evaluate the impact of corrective action and gives feedback to the project manager.

The managerial organization of this process is shown in Figure 11.1.

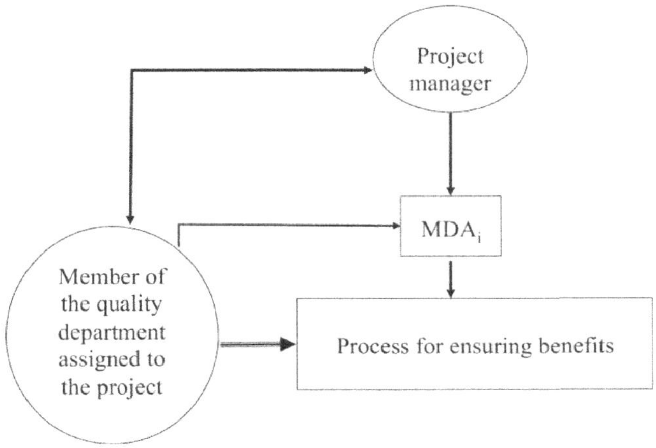

Figure 11.1. *Managerial organization of a processor for ensuring benefits*

COMMENT ON FIGURE 11.1.– MDAi denotes the member of the management, within the company, responsible for relations with a specific class of stakeholders (human resources management, financial service management, marketing management, purchasing management, administrative management, legal department, etc.).

11.6. Conclusion

The uncertainties inherent in any innovation process make it necessary to constantly oversee the evolution of risks of not obtaining the compensation that the stakeholders are expected to provide, which are often key to the success of the innovation.

In general, this supervision is relatively natural, and the stakeholders send signals which are easy to understand when they believe that the evolutions in the relations with the project team, or company, do not comply with the clauses in the contract as they had been initially perceived.

However, it is a more sensitive process to quickly carry out a diagnosis intended for identifying and quantifying the corrective action necessary. An innovation process can fail if a stakeholder who should provide essential resources decides to withdraw their contribution from the project.

Sometimes, the company has no responsibility for exchanges being stopped with one of its stakeholders. There are unpredictable evolutions in the stakeholder's quality comburant, or in its company environment, which are at the source of such situations. A change in policy, for example, can lead to public subsidies which were initially negotiated being lost.

We often share the following hypothesis in representation models of management: "all other things being equal". In the field of innovation, this is almost impossible. It is almost always the case of an "open field", hence the importance of correctly identifying the driver of socioeconomic exchanges in the innovation process from the beginning.

The balancing act of an innovation process means that relations with these stakeholders resemble the proverbial house of cards.

In new forms of management, it is common to talk about "agility". In an innovation process, feature is very much at the forefront.

12

The Role of the Quality Department

12.1. Introduction

This chapter discusses a sensitive subject. It is often difficult to make "research and development" structures and "start-ups" accept a representative from a quality department.

Even recently in large companies, people believe that quality departments are simply intended to oversee the production of products and services, ensuring an identical reproduction (otherwise known as quality consistency). Helping members of an innovation process to define a mobile "quality target" does not initially seem to fall within the remit of the quality department. In small businesses that only carry out innovation processes now and then, the stakes are so high that directors are more willing to request the quality department to assist the process. However, it is unfortunately most often the case that these departments do not have the skills necessary to fully play their part. Conflicts can emerge between creators and the quality department, which is considered as a source of constraints incompatible with creative research.

The developments in the previous chapters, however, show the need for specialized assistance in producing and using quality within an innovation process.

This specific assistance for project managers is based on knowledge and skills similar to those used in the quality department. This is why we advocate the contribution of members of this department, provided that they have acquired a specific expertise which we will attempt to describe in this chapter.

12.2. Positioning a quality department within an innovation process

First, it is a case of assistance, which means that at no point the member of the quality department in the project team is responsible for the quality produced and used by the innovation process to obtain the desired success.

In general, his role is limited to:

> *"Helping the project manager and the various members of his team to produce and use quality in the most cost-effective way, as part of their innovation process".*

We previously indicated that this objective could be expressed with more precision in the following ways:

– to promote the most cost-effective use of stakeholders' underlying competitive quality judgments during the innovation process;

– to help design, partially lead and ensure the maintenance of systems regulating the production of competitive quality fuel (CQF) released by the innovation process;

– to contribute to the attractiveness and "capital confidence" of the company for stakeholders by proving the effectiveness of the competitive quality process (CQP) integrated into the innovation process.

These objectives involve five lines of action, which can sometimes be in opposition to one another:

– to assist the members who carry out the stages of the CQP in the innovation process;

– to maintain an acceptable criticality for risks of not achieving the performance desired of the CQP;

– to contribute to the use of opportunities for improving the performance;

– at the request of the project manager, to communicate with stakeholders to prove the credibility of information provided which make up the competitive quality fuel;

– to track the key facts which can provide proof of compliance with the demands formulated by certain stakeholders, or which can be used in the future to benefit from feedback in order to produce the competitive quality of future innovation processes.

Taking advantage of opportunities can increase the criticality of risks, and to keep this at acceptable levels, the member of the quality department must carry out a supervisory role, which could be perceived as an intrusive inspection incompatible with the trust necessary in the assistance dynamic.

These members of the quality department will henceforth be referred to as “quality specialists” to simplify the explanations that follow.

The implementation of these two lines of action often seems to those team members working with the quality specialists, as being contradictory. It is difficult to offer assistance, supervise and regulate, and be accountable to the upper management level.

The quality specialist must not try to impose regulations or methodological tools. The project manager will take decisions regarding opportunities to conform to certain standards. The quality specialist is simply required to inform the team of their existence and help them: to use them and inspire confidence in their implementation for those stakeholders who demanded them.

Respecting regulatory demands in terms of health, for example, is the responsibility of the project manager. To conform to them, he requests the assistance of a specialist in regulatory affairs, who is often a member of the company's quality department.

The quality specialist will base his actions on a model of the CQP which may be the physical version of the generic model discussed in the previous chapters.

This extremely technical model is one of his work tools. He is not required to share this with the other members of the project team. Each of his interventions can be justified, if necessary, to these members by partially explaining the logical aspects of common sense which make up this model, and by using the distinct and specific language of the project in question. It is not up to the members of the project team to adapt their language to that of the quality specialist, but on the contrary, the quality specialist must translate his technical language into the language normally used in the project, to make the CQP as factual and natural as possible.

Another major challenge in the quality specialist's work is the choice of moments to suggest and assist the work of other team members contributing to various areas of the CQP. We offer some guidelines to ensure that the work of the quality specialist is correctly incorporated to the timing of the project.

Furthermore, given that this is a "one-shot" process, the same can be said for the CQP integrated into it. There are no "standard" actions that can be systematically applied. There are only specific actions that can be applied at a given time which are incorporated as naturally as possible, and in the most efficient way, in the general innovation process schedule. These specific actions may be adaptations of generic models.

The members of the quality team must understand at all times the importance and goals of the quality specialist's recommendations.

Obviously, the quality specialist must use the feedback received during previous innovation processes as well as possible, while avoiding making them into rules to be imposed.

He simply passes on a "quality culture" through events experienced during various stages of the CQP integrated into the innovation process. It is clear that his work is made easier if the company already has a strong culture based on quality.

One of the fundamental managerial questions asked is the following:

> *Must the quality specialist depend on the company's quality department?*

Our experience would encourage us to respond with the affirmative.

This availability is provided as part of a "client–provider"-type contractual relationship between the management of the "research and development" department and the quality department.

The quality specialist(s) made available can be assigned for a very long period to the research and development department so as to acquire the specific expertise necessary to carry out his objectives as part of the innovation process.

The contract agreed between the departments of research and development, and quality specifies how the relations should take place between the staff in the quality department and those assigned to research and development, which is involved in the innovation process.

It also sets out the objectives that this staff must meet and how they are to do so.

The contract finally specifies how relations are to be maintained between these staff members and the quality department.

By using the definitions given in Chapter 1, which discuss the managerial structure of the company's human resources, the staff of the quality department made available to the research and development department are part of the pyramid-shaped managerial structure of the quality department and the horizontal organization of the processor which carries out the innovation process.

When these staff members are assigned to an innovation project, they are responsible for helping the project manager to optimize the performance of the CQP integrated into the innovation process.

These staff members take orders from the project manager. They provide advice, oversee certain actions taking place during the CQP or carry out some actions themselves when it demands particular skills that the project team do not possess, or when they incompatible with the necessary availability of these team members to carry out the production of added value, it the best possible conditions, which is at the core of their objectives.

To be more precise, each objective of one of these members can be defined by a set of specifications indicating:

– the technical function(s) of the CQP in the innovation process contributed by the quality specialist;

– the resources that it provides to carry out its objective;

– the constraints to which it is subject (structure and operation of the processor that carries out the innovation process);

– the nature of the support given to other members of the project team.

The company's quality department is therefore simply required to provide the project manager with the human resources that have the skills necessary to carry out their objectives within the innovation process.

12.3. The specificities of the quality specialist's objectives

One of the main lines of action which characterizes the role of the quality specialist is often called "dynamic regulation".

It is carried out in four ways:

– participative regulation to define mobile targets of the CQP in order to reap all the benefits possible;

– participative regulation, through anticipation, of the risks of not meeting these targets and estimating their criticality;

– regulation of real risks that appear during the actions of the CQP;

– regulation of actions that correct deviations from the performance predicted in the CQP.

Each type of regulation is carried by triggering nine different actions that can be described in the following ways:

– observing a situation;

– interpreting information obtained;

– envisaging, or encouraging others to envisage, the change to be carried out in the observed situation;

– suggesting a decision be taken on carrying out this change;

– assisting, or participating, in making this change;

– comparing the results of the change with the objectives;

– estimating the probable effects of this change;

– presenting an evaluation of the work carried out;

– recording the key events.

The first two stages in this type of regulation, as well as the final three (text not in italics) do not need to interfere with the other activities of the project. They should not disrupt the progress of the project. Only the stages in italics are visible to the project team.

The team members whom the quality specialist has frequent communication with are:

– the project manager;

– the departments specialized in communication with each class of stakeholders;

– the other members of the project team;

– the stakeholders in the project.

These specialized departments, when they exist, are normally the following:

– finance department (relations with investors);

– marketing department (relations with future clients of the products or services which incorporate the innovation);

– purchasing department (relations with providers for the project and the future business unit (BU));

– human resources department (relations with the members of the project team and with members of the future BU which will produce the products or services incorporating the innovation);

– departments responsible for relations with social regulatory bodies (environment, safety, legal, etc.) .

This regulation is complex because: the company's relations with its environment are complicated; human beings play a key role in the production of quality within the innovation process, since this role is based on individual and collective attitudes and behavior; and because quality is a concept that entails a significant level of irrationality.

The quality specialist can also assist the project team members, or these departments, when they carry out exchanges with stakeholders, to help them to be more effective in "selling" the competitive quality fuel that is intended for them. This is the second type of intervention which the quality specialist can make. He becomes a kind of ambassador of the project for some classes of stakeholders. He is one of the project's CQF emitters providing information intended to prove the capacity of the project to comply with the commitments made to certain stakeholders, and to the standards imposed by social regulations.

The quality specialist operates differently depending on the position of the stage in the CQP, within creative research, impact studies, feasibility studies, organization or decision-making.

12.4. The objectives of the quality specialist in creative research

In the various stages of the creative research, the quality specialist mainly carries out actions to regulate the definitions of mobile targets of the CQP in order to reap maximum benefits, in the ways previously described.

He may also, occasionally but not too frequently, carry out actions to regulate by anticipating the risks of not meeting these targets.

The previous chapters help to bring together the various stages in the CQP which operate during creative research in the following way.

	CQT	Marketing quality	CQF	Specifications of emitters	Qualification of emitters	Validation of marketing	Supervision of capability	Treating non-compliance	Supervision of perceived quality	Ensuring benefits	Records of improvements
Bringing about the emergence of an idea that may provide a response to a need of society	C										C

Creating experimental drafts that provide a more accurate representation of the idea	C	C	C I	I							C I
Progressively designing drafts to accurately represent the configuration of the innovation	C	C	C I	C I							C I
Progressively designing drafts that give a representation of future products and services to incorporate the innovation, by describing in particular, the future "interfaces" of the innovation with other components of these products and services	C	C	C I P	C I P				C			C

Box 12.1. *Creative research in the CQP*

COMMENT ON BOX 12.1.–

Class (C) of stakeholders includes:

– the clients of future products or services incorporating the innovation (the innovation's potential market).

Class (I) includes:

– the stakeholders who produce the innovation (the project team);

– the stakeholders, outside the project team, who provide certain components for the innovation, or resources, which are then transformed by the innovation team to produce the innovation (project providers);

– the stakeholders who invest in the project hoping to reap the benefits from the economic utilization of the innovation (project investors);

Class (P) includes:

– the stakeholders in the BU who produce certain components for the future products or services which will incorporate the innovation;

– the stakeholders outside the BU who will provide certain components of these products or services, or resources which will then be transformed by the company to produce these products or services (providers);

– the stakeholders who will invest in the BU hoping to reap the benefits of producing these products or services (investors).

To provide more detail on these objectives, we will review the various stages of creative research given in Box 12.1.

12.4.1. *The quality specialist's interventions in the emergence of the idea*

The various components of the CQP which are involved in the emergence of the idea are summarized in the box below.

	CQT	Marketing quality	CQF specifications	Specifications of emitters	Qualification of emitters	Validation of marketing	Supervision of capability	Treating non-compliance	Supervision of perceived quality	Ensuring benefits	Records of improvements
Bringing about the emergence of an idea which may provide a response to a need of society.	C										C

Box 12.2. *The CQP during the idea emergence stages*

The only stakeholders from class (C) involved are the end consumers of the products or services.

At this stage of the innovation process, the CQP only intervenes to start targeting future end consumers of the products or services that will incorporate the innovation, and to envisage the general guidelines of future competitive quality tactic (CQT) to implement in exchanges with these future economic stakeholders.

Most often, the quality specialist who must assist the project has not yet been nominated at this stage of creative research.

If he is already present, his role is simply to suggest to the project manager this framework for the CQT by providing him with methodological assistance if necessary.

The triggers of the idea are often observations of events that take place during the consumption of products or services similar to those that may incorporate the innovation. Consequently, the future users of the innovation are easily targeted and the initial interest in the new idea is instinctively linked to a predictive vision of the guidelines for the CQT.

However, this predictive vision may not always be explicit at this stage of the innovation process. The role of the quality specialist, if he is already present, is to facilitate its expression and make a record of it.

12.4.2. *The interventions of the quality specialist in the creation of experimental drafts*

The various components of the CQP integrated in the idea emergence stage are summarized in the following box.

	CQT	Marketing quality	CQF specifications	Specifications of emitters	Qualification of emitters	Validation of marketing	Supervision of capability	Treating non-compliance	Supervision of perceived quality	Ensuring benefits	Records of improvements
Creating experimental drafts that provide a more accurate representation of the idea	C	C	C I	C I							

Box 12.3. *The DQC during the experimental draft creation stages*

This stage of the innovation process, which can be carried out in a number of sequences, must include the components of the CQP which concern several categories of stakeholders.

Those who are the most affected are:

– the end consumers of the products or services incorporating the innovation;

– the bodies financing the project;

– the members of the project team.

The stages of the CQP which must be partially taken into account at this stage are the following:

– continuing the design of CQT which affects the relations with future users of the products or services incorporating the innovation;

– producing the initial scenarios for the process of producing and releasing competitive quality fuel for the future end consumers of the products or services incorporating the innovation;

– identifying the basis of primary information which will then be used to design various CQF, directed at:

- future end consumers of products or services incorporating the innovation;

- bodies that finance the project;

- various members of the project;

– identifying the basis of future CQF emitters.

For the last two components of the CQP, it is a case of rapidly scanning all stakeholders who, for one reason or another, will be involved in the project, either financing it or participating in the production of its added value, and identifying the types of information which are likely to prompt perceptions whereby their expectations are met.

Great attention must be paid, at this stage of the innovation process, not to stifle the creation by placing too many constraints which may be imposed by various project stakeholders.

These drafts must above all reflect the attractiveness that must be provided by the innovation for future clients and that must be at the heart of their competitive quality judgments.

The role of the quality specialist is to attract the attention of the project manager to the importance of starting to envisage the initial scenarios for the process of designing and releasing the CQF which will be aimed at future end consumers of the products or services incorporating the innovation, and to lay, in the construction of drafts, the foundations for various competitive quality fuel which will be aimed at these stakeholders.

He may suggest some simple methodological tools.

He must also attempt to identify the foundations for the primary information which will then be used to design various CQF aimed at the bodies financing the project and to the various future members of the project team. In addition, he must attempt to identify the basis of the future CQF emitters.

The seemingly natural lines of action which these components of the CQP rely on actually cause problems for the designers of the initial experimental drafts because they generally do not have enough information on these stakeholders' quality comburants.

At this stage of the innovation process, confidentiality and the necessary freedom in the creation process mean that not many members should be involved in this type of action.

The quality specialist must therefore be able to accept very partial implementations of tasks for the CQP, even if this means taking an inventory and putting to one-side future sources of progress.

12.4.3. *Interventions of the quality specialist in the creation of drafts of the innovation*

This stage of creative research consists of progressively designing drafts which specify the configuration of the innovation.

The various components of the CQP involved in this stage are summarized in the box below.

	CQT	Marketing quality	CQF specifications	Specifications of emitters	Qualification of emitters	Validation of marketing	Supervision of capability	Treating non-compliance	Supervision of perceived quality	Ensuring benefits	Records of improvements
Progressively designing drafts to accurately represent the configuration of the innovation	C	C	C I	C I							

Box 12.4. *The CQP during the stages of creating drafts of the innovation*

This stage of creative research helps to move the CQP forward in terms of:

– the end consumers of the products or services incorporating the innovation;

– the bodies financing the innovation project;

– the contributors to the project who will be the future users of the products or services, or the future producers of the productions system of these products or services;

– external members providing human, material and information-based resources;

– the members of the project team;

– the departments in the company responsible for the protection of: the environment, people and property.

The project manager must create, or design, the products or services which are to be offered to various stakeholders in this list, other than the future end consumers of the products or services incorporating the innovation. He must take advantage of this stage of creative research to start characterizing the CQF integrated into these products or services.

For instance, these products and services may take the form of: the business plan offered to the bodies financing the project; or draft contracts, including the technical specifications which will apply to external members providing human, material or information-based resources; or plans for tests intended to comply with various regulations which apply to the innovation, etc.

The various actions to carry out for the CQP during this stage of creative research can be described as follows.

1) For the CQT, only those stakeholders using the products or services incorporating the innovation are concerned. The aspects of the CQT to be examined at this stage for each of the categories of future users of the innovation can be described as follows:

i) identifying the general expectations, not satisfied by the competition, to which future users attach a significant level of importance, and which the project is able to satisfy more fully;

ii) identifying the main behavioral mechanisms of these stakeholders who are likely to produce their quality perceptions to the responses to expectations, and which will cause them to express a competitive quality judgment.

At this stage, the role of the quality specialist is:

i) to suggest to the project manager to continue designing the CQT to adopt the use of the products or services incorporating the innovation in exchanges undertaken with stakeholders;

ii) to assist the project manager by working with the company's marketing department, if it exists, to ensure that the information to help design the CQT is collected at this stage. If this department does not exist, he can provide methodological tools to help certain members of the project team to carry out such actions;

He provides feedback to the project manager regarding the correct execution of these actions.

2) The components of the creation of scenarios for the process of designing and releasing CQF aimed at these members to be examined at this stage can be described as follows:

i) identifying the events to integrate into the creative research stage of the innovation process which will contribute to the design of the CQF aimed at future users, in compliance with the provisions of the CQT;

ii) identifying the events likely to be observed which will describe the impact of this competitive quality fuel on their competitive quality perceptions;

iii) identifying the observable events which help to form a judgment on the compliance of the compensatory resources which will be provided by the future users with the expectations of the company;

iv) describing a pre-defined flow diagram of these events and their position in the general schedule of the stage of creative research;

iv) to move this process forward during this stage of the innovation process, complying with the provisions of the CQT.

The role of the quality specialist is to suggest to the project manager to design scenarios of the processes for producing and releasing CQF aimed at these members.

He carries out these tasks with the assistance of the project manager and the marketing department. He may suggest appropriate methodological tools.

3) The continuation, during the creation of the drafts, of laying the foundations of various CQF of the products and services, and their emitters, which are aimed at:

i) end consumers of products or services incorporating the innovation;

ii) the bodies financing the innovation project;

iii) the contributors to the project who will be the future users of the products or services, or the future producers of the productions system of these products or services;

iv) external members providing human-, material- and information-based resources;

v) the members of the project team;

vi) the departments in the company responsible for the protection of: the environment, people and property.

This process consists of:

i) reminding various members involved in these designs of the provisions already established in the CQT which only apply to the competitive quality perceptions to be prompted from these classes of stakeholders;

ii) recalling the aspects of the scenarios of various marketing quality elements which have already been established;

iii) progressively envisaging the nature of the information which may be released by these products or services in order to prompt these perceptions of the provisions and scenarios;

iv) positioning the release of information in these scenarios;

v) locating various possible vehicles for the release of this information;

vi) starting to envisage certain specifications of this information so that their release gives the best possible chance of prompting the competitive quality perceptions predicted in the CQT.

The competitive quality perceptions targeted by these drafts essentially:

i) hope to satisfy expectation, or demands of the various stakeholders discussed previously;

ii) trust in the capacity of the innovation process to produce these products or services for these stakeholders, whose CQF complies with the information already given.

At this stage of creative research, the quality specialist can ask the project manager to ensure that the specifications of the various CQF and their emitters can be refined. These are CQFs of that offered to end consumers of the products or services incorporating the innovation, and to all stakeholders in the innovation project.

He can suggest to the project manager to submit these drafts to representative of the finance, marketing, purchasing, human resources and other departments responsible for social regulation, which are assigned to the project, in order to give an opinion on the relevance of the CQF and their emitters which are integrated in these drafts.

He may also, as before, propose specific methodological tools and lead their implementation.

He may even suggest ideas to change drafts so as to increase their relevance, or the visibility of the fuel and their emitters.

12.4.4. *The interventions of the quality specialist for the creation of drafts of future products or services*

The various components of the CQP which are involved in the stage where the idea emerges are summarized in the box below.

	CQT	Marketing quality	CQF specifications	Specifications of emitters	Qualification of emitters	Validation of marketing	Supervision of capability	Treating non-compliance	Supervision of perceived quality	Ensuring benefits	Records of improvements
Progressively designing drafts which give a representation of future products and services to incorporate the innovation, by describing, in particular, the future "interfaces" of the innovation with other components of these products and services	C	C	C I P	C I P				C			C

Box 12.5. *The CQP during the stages of draft creation for future products or services*

CQP activities during the progressive design of these drafts which concern various stakeholders using the products or services incorporating the innovation (C) can be modeled in the following ways:

1) During this period, the relevance of the CQT to be applied to the exchanges with each category of this class of stakeholders can be improved, specifying certain general characteristics to be included in the CQF of products or services incorporating the innovation.

The role of the quality specialist is the same as described during the previous stage of creative research:

i) to suggest to the project manager to continue designing the CQT to adopt the use of the products or services incorporating the innovation in exchanges undertaken with stakeholders, and which concern certain general characteristics to include in the CQF of the products or services incorporating the innovation;

ii) to assist the project manager by working with the company's marketing department, if it exists, to ensure that the information to help create these drafts is collected at this stage. If this department does not exist, he can provide methodological tools to help certain members of the project team to carry out such actions.

He provides feedback to the project manager regarding the correct execution of these actions.

2) The creation of marketing quality likely to take place in the exchanges with each of these types of stakeholders is continued by:

i) identifying events likely to take place during the design of drafts of the products or services, and the presentation of drafts to the various categories of stakeholders who will use them, and:

– who will release information which stimulate their competitive quality perceptions, complying with CQT provisions.

– who will help observe information describing:

- how the CQF information will have an impact on their competitive quality perceptions;

- how these quality perceptions will affect the quality of compensatory resources provided by these stakeholders to the company;

- how these stakeholders affect the image or reputation of the company.

ii) describing a pre-established and organized flow diagram of each of these events;

iii) presenting this diagram to some of the stakeholders in order to obtain their reactions;

iv) interpreting their reactions in terms of opportunities or risks regarding the effects of their competitive quality perceptions, the provision of compensation to the future BU and their contributions to the image of the project and reputation of the company;

v) if necessary, adapting the initial diagram to ensure that the CQT provisions are better complied with.

At this stage, all types of competitive quality perceptions of these stakeholders on the use of the products or services incorporating the innovation should be targeted.

We should recall that these can be summarized in the following ways:

– perceptions of expectations concern:

- the satisfaction of expectation by certain components of the production of the innovation process, its results and the effects they have;

- the ease of access to use such components;

- the process of making efforts cost effective in order to understand them, obtain them and use them.

– perceptions of trust in the capacity of the innovation process to comply with the objectives declared.

Once again, the role of the quality specialist is:

– to suggest to the project manager to continue creating competitive marketing quality which will take place in the exchanges with each of these types of stakeholders;

– to lead the completion of these tasks with the assistance of the project manager and the marketing department;

– to propose appropriate methodological tools.

3) The description of CQF specifications released as the events of previous scenarios are carried out is moved forward, by continuing to:

i) locate the various possible vehicles for these events (processors);

ii) design specifications of information produced by these events which have the best chances of prompting these perceptions.

As before, he may also propose specific methodological tools and lead their implementation.

He may even suggest ideas to change the drafts so as to increase their relevance, or the visibility of this fuel and its emitters.

4) The design of the CQF emitters is also moved forward, by continuing to refine:

i) the location and characteristics of the primary information which could be used to release CQF;

ii) the scenarios of how this primary information is transformed into quality fuel which complies with their specification;

iii) the design of processors which obtain this primary information, their transformation into quality fuel and the communication of quality fuel to stakeholders.

The role of the quality specialist is to:

i) suggest to the project manager to continue with the design of CQF emitters with the methods described above;

ii) help project team members carry out these tasks;

iii) evaluate the work carried out.

5) The guidelines for checking for aspects of non-compliance to CQF specification can start to be defined.

The role of the quality specialist is to fully complete this task under the responsibility of the project manager.

The design of these drafts must also promote the creation and release of the CQF of the products or services offered to some categories of the class of stakeholders in the innovation project (I), and the production of the products or services incorporating the innovation (P).

The various categories of stakeholders concerned in class (I) are the following:

i) bodies which finance the project;

ii) future users of the products or services incorporating the innovation, or future producers of the production system for products or service incorporating the innovation and who participate in the project;

iii) departments in the company responsible for the protection of the environment, people and property.

The various categories of stakeholders concerned in class (P) are the following:

i) bodies which finance the design of the production system for products or system incorporating the innovation;

ii) purchasers of products or services which include the innovation;

iii) members inside or outside the company who supply the production system with human-, material- or information-based resources;

iv) members inside the production system contributing to the production of added value.

For each of these categories of stakeholders of the specific products or services, the project manager proposes the compensatory resources which are required of them.

The involvement of the creative research stage in the design and release of the CQF of the products or services consists of:

i) completing the CQT to be applied to exchanges with these various stakeholders during the design of these drafts;

ii) contributing to the process of envisaging the nature of the information which can be released during the design of these drafts to prompt competitive quality perceptions compliant with the guidelines in the CQT;

iii) positioning the release of this information in the competitive marketing quality scenarios;

iv) locating the various possible vehicles for this information;

v) contributing to the design of the specifications of this information to have the best chances of prompting the desired perceptions when they are released.

The role of the quality specialist is to ensure, during the creation of the drafts of future products or services, the design of the CQF integrated into the products of services proposed to stakeholders previously targeted in class (I) and (P), while complying with the provisions of the CQT.

He continues to propose methodological tools, and ensures that the company's departments which are regularly in contact with these stakeholders are involved in these actions.

Once again, it is necessary to find the right compromise between the result of the creative research and the start of the real design process of the products or services. It is important to remain at the drafting stage, which means that the specifications of the CQF concerning stakeholders' use of future products or services incorporating the innovation also remain imprecise. The specification of the CQF, and its emitters, for products or services aimed at other classes of stakeholders must be more precise, even if they may need to be modified or supplemented by impact or feasibility studies.

12.5. The missions of the quality specialist in impact studies on competitive quality fuel

It is at this stage of innovation process that the role of the quality specialist is the most important.

The various stages of the CQP during the impact studies are summarized in the following box.

	CQT	Marketing quality	CQF specifications	Specifications of emitters	Qualification of emitters	Validation of marketing	Supervision of capability	Treating non-compliance	Supervision of perceived quality	Ensuring benefits	Records of improvements
Briefly characterizing products or services which could benefit from the innovation		C	I	I						I	C I
Identifying possible clients for these future products or services.	C		I	I					C	C	C I
Characterizing future uses for products or services which would incorporate the innovation	C	C	C I	C I	C	C	C		C	C	C I
Conducting prospective studies on possible socioeconomic uses for the innovation, and its impacts on the company and its position	C I P	C I P	C I P	C I P	C	C				C	C I P
Conducting tests on the drafts of future products or services, at various steps in their production, on future clients and on all future stakeholders in the production system for the products and services	C P	C P	C I P	C I P			C		C P	C P	C P

Box 12.6. *The CQP in impact studies*

COMMENT ON BOX 12.6.–

Class (C) of stakeholders includes:

– the clients of future products or services incorporating the innovation (the innovation's potential market).

Class (I) includes:

– the stakeholders who produce the innovation (the project team);

– the stakeholders, outside the project team, who provide certain components for the innovation, or resources, which are then transformed by the innovation team to produce the innovation (project providers);

– the stakeholders who invest in the project hoping to reap the benefits from the economic utilization of the innovation (project investors).

Class (P) includes:

– the stakeholders in the BU who produce certain components for the future products or services which will incorporate the innovation;

– the stakeholders outside the BU who will provide certain components of these products or services, or resources which will then be transformed by the company to produce these products or services (providers);

– the stakeholders who will invest in the BU hoping to reap the benefits of producing these products or services (investors).

To provide more detail on these missions, the various stages of the impact studies identified in Box 12.6 will be individually reviewed.

12.5.1. *The role of the quality specialist in the characterization of future products or services*

This first stage of impact studies is intended to identify the existing products or services whose added value would be increased if they benefited from the innovation. Sometimes, these do not exist and the innovation itself is the future product or service to be designed. In this case, a functional analysis is generally carried out which helps to give a picture of the initial configuration of this product or service. This stage is situated very early in the innovation process.

The role of the quality specialist is limited to regulating the definitions of the CQP's mobile targets in order to reap maximum benefits.

The various components of the CQP involved in this stage are summarized in the following table.

	CQT	Marketing quality	CQF specifications	Specifications of emitters	Qualification of emitters	Validation of marketing	Supervision of capability	Treating non-compliance	Supervision of perceived quality	Ensuring benefits	Records of improvements
Briefly characterizing products or services which could benefit from the innovation.	C		C I	C I							C I

Box 12.7. *The CQP during the characterization stage of future products or services*

The components of the CQP involved in this stage of impact studies can be characterized as:

– a continued design of the CQT to be applied: in exchanges with those speculating on the future use of the products or services incorporating the innovation, and in the future exchanges with the end consumers of these products or services;

– a reflection on the CQF which could prompt competitive quality judgments from these parties and members of the project team.

These reflections on the CQT are limited to:

– describing the macro-characteristics of the demands or expectations of these parties in relation to the products or services that the company could offer them by integrating the innovation into these products or services;

– briefly identifying the CQT of the main competitors which may disrupt exchanges with these parties;

– attempting to identify the expectations, which are not satisfied by the competitors, to which these parties attach great importance, and that the project is able to satisfy better;

– observing the behavioral mechanisms of these parties which form the basis of their quality perceptions.

The reflection on the CQF to be envisaged mainly discusses:

– the nature of the information which could be released during this stage of the impact studies to prompt quality perceptions from members of the project team, and those responsible for investment in the company;

– the location of various possible vehicles for this information.

The missions of the quality specialist are the same as those described previously at the idea emergence stage.

At this stage of impact studies, the quality specialist may not yet be assigned to the project. If he has been assigned, his activity is limited to observing the work of the initial members of the project team, and suggesting to the project manager to start addressing these two components of the CQP.

12.5.2. *The role of the quality specialist in identifying future clients*

This second stage of the impact studies is intended to identify the economic parties who are likely to use, for one reason or another, the products or services incorporating the innovation, and how what uses these are.

When a marketing department exists, it is the main party in carrying out this stage. The role of the quality specialist is again limited to regulating the definitions of mobile targets of the CQP in order to reap the maximum benefits.

The various components of the CQP involved in this stage are summarized by the following box:

	CQT	Marketing quality	CQF specifications	Specifications of emitters	Qualification of emitters	Validation of marketing	Supervision of capability	Treating non-compliance	Supervision of perceived quality	Ensuring benefits	Records of improvements
Identifying possible clients for these future products or services	C		I	I					C	C I	C I

Box 12.8. *The CQP during the future client identification stage*

The components of the CQP integrated into this stage of impact studies, which concern all categories of economic parties likely to use the innovation, can be characterized by:

– the continued design of CQT which can be applied in exchanges with these parties;

– an initial acknowledgement of the guidelines for helping to carry out the supervision of these parties' perceived quality;

– an initial estimate of the possible benefits the company can expect, which is obtained by developing exchanges with these parties.

The components of the CQP involved in this impact study stage, concerning the members of the project team, can be characterized as:

– the continued definition of CQF specifications and its emitters;

– the partial creation of the set of specifications for the products or services expected from the parties within the project.

In the CQT, it is simply a question of:

– describing the macro-characteristics of the demands, or expectations of these potential future users of the innovation that it could satisfy;

– identifying the CQT of the main competitors who may disrupt exchanges with these stakeholders;

– among these expectations, identifying those which are insufficiently satisfied by the competition, to which future users attach great importance, which the innovation is able to better satisfy;

– starting to understand the behavioral mechanisms of these parties which are the basis of their quality perceptions;

– starting to envisage the impact of the exchanges on the company which could take place with these parties due to the innovation.

In terms of acknowledging the guidelines for overseeing perceived quality, it is simply a question of:

– starting to identify certain information that these parties may release and which describe their competitive quality perceptions.

As regards an initial estimate of the possible benefits expected by the company, which is obtained through the development of exchanges with them, this process is limited to defining an outline of the specifications of the compensation provided following the use of the innovation which could be requested from each category of stakeholder.

The role of the quality specialist is to suggest to the project manager to take advantage of the identification of future economic parties who are likely to use the innovation to continue to configure these specific components of the CQP. When the project manager gives the marketing department the operational responsibility of these impact studies, the quality specialist assists by trying to elicit information which can provide guidelines regarding how to prompt these parties' future competitive quality perceptions.

The continued definition of certain specifications of the CQF intended for members of the project team and its emitters is limited to detecting some primary information likely to be used to design the CQF.

The quality specialist carries out this mission independently of the fringes of the project.

The partial design of the specification of the products or services expected by these parties within the project is the responsibility of the project manager.

12.5.3. *The role of the quality specialist in the characterization of future uses of the innovation*

This third stage of the impact studies is intended to envisage what the future uses of the innovation could be through the intermediary of the products or services which will integrate it.

When a marketing department exists, it is also the main protagonist in this stage. The role of the quality specialist is again to ensure a regulation to obtain guidelines in these descriptions in order to develop the quality of the innovation as best as possible.

The various components of the CQP which are involved in this stage are summarized in the following box.

	CQT	Marketing quality	CQF specifications	Specifications of emitters	Qualification of emitters	Validation of marketing	Supervision of capability	Treating non-compliance	Supervision of perceived quality	Ensuring benefits	Records of improvements
Characterizing future uses for products or services which would incorporate the innovation	C	C	C I	C I					C		C I

Box 12.9. *The CQP during the stage of characterizing the future uses of the innovation*

The stakeholders targeted as a priority by the CQP integrated into this stage of the innovation process are those which will use the products or services incorporating the innovation. This characterization of the future uses of the innovation will also produce primary information which will be used to create the CQF of project funders and members of the project team.

Components of the CQP involved in this stage, which affect the future users of the innovation can be characterized as:

– the continued design of the CQT;

– the continued design of scenarios which depict the events which could take place during exchanges with these parties to apply to the CQT;

– the continued identification and characterization of information which could make up the CQF produced by the innovation in the products or services incorporating it;

– the characterization of the emitters of this information;

– the continued identification of guidelines helping to oversee information emitters, by these parties, which express their perceptions regarding the quality of the innovation.

In terms of competitive quality marketing, it is a case of:

– identifying events which are likely to be observed during the use of the innovation, and which describe how the information of the CQF produced by the innovation will have an impact on their competitive quality perceptions of these stakeholders;

– describing a predefined and organized initial flow diagram for each of these events.

The role of the quality specialist is to guide the research among information on the characterization of the future uses of the innovation, which helps to highlight the observable events of the CQF impacts produced by the innovation on its future users on their competitive quality perceptions. This work is often a coproduction with the marketing department when it exists.

In terms of characterizing information which could contribute to the primary information of the CQF of the future products or services incorporating the innovation, and that of their emitters, it is a case of:

– continuing to envisage the nature of the information which can be released to prompt these competitive quality perceptions during the various stages of the innovation process;

– positioning the release of this information in the competitive marketing quality scenarios describing some events which are likely: to be produced during the project, to be involved in the communication process with these parties or coproduced with them, and which release the information that stimulate their competitive quality perceptions;

– locating the possible emitters of this information;

– continuing the design of specifications of this information in order to have the best chances of prompting these perceptions when it is released;

– highlighting the characteristics of this primary information which are supplied to the emitters;

– starting to design scenarios of the process for transforming this primary information into quality fuel, in compliance with their specifications.

The role of the quality specialist then becomes essential. He leads the tasks which carry out these actions. The project manager must request members of the project team to collaborate on these tasks.

In terms of continuing to identify the guidelines for overseeing the release of this information, by these parties, which express their perceptions on the quality of the innovation, it is up to the quality specialist, as before, to carry out this task independently of the fringes of the project.

For the components of the CQP integrated into this stage, which are addressed at project funders and members of the project team, it is simply a case of using some of the data from the studies to contribute to the composition of the CQF of the products or services which will be aimed at them, to encourage them to supply the project with the resources it needs.

The role of the quality specialist is to highlight the information produced by these studies which will be used to design the CQF aimed at project funders and members of the project team.

12.5.4. *The role of the quality specialist in carrying out prospective studies*

This fourth stage of the impact studies is intended to envisage the general impact the innovation will have on the configuration of the company and its strategic areas of activity.

This macroeconomic projection into the future is the responsibility of the company management, which is assisted by the project promoter or project manager.

The role of the quality specialist is to suggest to the project manager to use these general studies to obtain guidelines to develop the quality of the innovation as best as possible.

The various components of the CQP which are involved in this stage are summarized in the following box.

	CQT	Marketing quality	CQF specifications	Specifications of emitters	Qualification of emitters	Validation of marketing	Supervision of capability	Treating non-compliance	Supervision of perceived quality	Ensuring benefits	Records of improvements
Conducting prospective studies on possible socioeconomic uses for the innovation, and its impacts on the company and its position	C I P	C I	C I	C I	C	C				C	C I P

Box 12.10. *The CQP during the stage of conducting prospective studies*

These studies are intended to anticipate the macroeconomic and macrosociological mechanisms which are likely to be produced by the innovation, and the predicted impacts on the company and its environment.

Chapter 1 discussed how any innovation can lead to an evolution in a company's expertise. The nature of this impact affects the qualifications and expertise of staff, new technology to be introduced into the company, the transformation of the operational infrastructure, the redirection of investments, and so on.

In the same way, it can modify the company's image or its offers in one of its strategic areas of activity. These changes in company image can be opportunities to grasp to modify the company policy.

For all these reasons, the prospective studies consist of studying the general possible impacts of the innovation on what we have called the driver of socioeconomic exchange in the company.

Chapter 1 introduced the concept of a driver of socioeconomic exchanges of an innovation. This concept can be generalized by characterizing the driver of socioeconomic exchanges in a company as:

– a system for producing value;

– the company;

– its stakeholders;

– neighboring parties who influence its exchanges with its stakeholders without having direct exchange relationships with it;

– a network of exchange flows of products, services or capital between the company and its stakeholders;

– a communications network and information flow with neighboring parties;

– exchange networks for information, products, services or capital between its stakeholders or between its stakeholders and neighboring parties.

By using this model, the prospective studies can be said to consist of envisaging the transformations that the innovation is likely to cause on the company system, on the exchange network of products, services or capital, between the company and its stakeholders, on the communication network and the flow information with neighboring parties, and finally, exchange networks of information, products, services or capital between its stakeholders or between stakeholders and its neighboring parties.

The main components of the CQP affected by this stage of impact studies are the designs of various CQT to implement in the relations with most classes of stakeholders in the innovation process.

It is a case of supplementing the CQT which have already been started in the previous stages in the innovation process.

This stage can also be used to review the other components of the CQP which have already been studied to adapt them according to the modifications made to the CQT.

The additions made to the CQT regard:

– the identification of the CQT of the main competitors which can disrupt exchanges with these stakeholders;

– the description of the macro-characteristics of demands or expectations, of these parties which are not satisfied by competitors, to which they attach great importance and which the project is able to better satisfy;

– the understanding of the main behavioral mechanisms which make up the stakeholders' quality fuel;

– the creation of a hierarchy of stakeholders' quality perceptions for the project to produce according to the importance of their impacts on their competitive quality judgments;

– the description of the main characteristics expected from services that the stakeholders should provide as compensation (expected benefits);

– the estimate of the capacity of stakeholders to produce them;

– the estimate of the perceived interest of stakeholders to provide the effort required in order to acquire the products or services offered by the project;

– the very general characterization of CQF of the products or services which should be offered to stakeholders in order to obtain their services.

During these studies, the quality specialist must suggest to the project manager to complete the design of the CQT to be adopted with all the stakeholders in the project, without getting involved in the operation of these processes.

However, he must propose modifications to other components of the CQP which has already been started in order to follow the guidelines set out in the CQT.

12.5.5. *The role of the quality specialist in carrying out tests on the drafts of future products or services*

This fifth stage of the impact studies is intended to test the drafts of the future products or services which will incorporate the innovation on a sample representing the future users and producers.

The various components of the CQP which are involved in this stage are summarized in the following box.

	CQT	Marketing quality	CQF specifications	Specifications of emitters	Qualification of emitters	Validation of marketing	Supervision of capability	Treating non-compliance	Supervision of perceived quality	Ensuring benefits	Records of improvements
Conducting tests on the drafts of future products or services, at various steps in their production, on future clients and all future stakeholders in the production system for the products and services	C P	C P	C I P	C I P			C		C P	C P	C I P

Box 12.11. *The CQP during the stage of testing the drafts of future products or services*

The components of the CQP which affect the users of future products or services are the following:

– verifying if the CQT to be introduced into exchanges which these classes of stakeholders contribute to the success of the innovation;

– observing how the CQF of drafts of future products or services, and their emitters, influence stakeholders' competitive quality perceptions;

– comparing these observations with scenarios of marketing quality which have already been drawn up;

– adapting these scenarios, specifications of this CQF and their emitters, where necessary;

Using this information to continue the design of the future system for supervising:

– the ability of the production system of products or services to continue with the design of the system;

– the way in which these stakeholders will perceive the quality of future products or services;

– completing the definitions of the compensatory resources which can be requested from the stakeholders;

– tracking the improvements which arise from these tests.

Some information provided by these tests must also help to supplement the same components of the CQP regarding the products or services proposed to most of the stakeholders in the production of the products or services incorporating the innovation.

The only components of the CQP which concern the stakeholders in the innovation project are those which contribute to designing the specifications for the CQF of products or services which are offered to them and the CQF emitters.

For the stakeholders who will use the future products or services, in terms of the CQT, it is a case of:

– validating the identification of expectations which are not fully satisfied by the competitors, to which they attach great importance and which the innovation is able to better satisfy;

– continuing the identification of future producers' expectations of these products or services to satisfy so that they agree to commit to the project;

– continuing to understand the behavioral mechanisms of these parties which are the basis of their quality perceptions;

– characterizing the competitive quality perceptions of the various categories of parties;

– continuing to estimate the impacts for the company of the exchanges which could take place with these parties due to the innovation;

– completing the very general characterization of the CQF of future products or services incorporating the innovation.

The role of the quality specialist is to attract the attention of the project manager to the possibility of improving the relevance of the CQT to be adopted in exchanges with users and producers of the future products or services, according to the information emerging from these tests.

In terms of competitive quality marketing, it is a case of:

– verifying whether the scenarios previously envisaged which describe how information of quality fuel of future products or services, and products and services offered to the future producers of these products or services will have an impact on the competitive quality perceptions and whether the contributions of these stakeholders are relevant to the success of the innovation;

– interpreting the reactions to the tests in terms of opportunities or risks regarding the effects on their competitive quality perceptions, on the provision of their compensatory resources to the project and on their contribution to the image of the project and the reputation of the company;

– adapting the initial flow diagram to comply with the provisions of the CQT.

The role of the quality specialist is to help the marketing department, if it exists, or the project manager, to use these tests to validate the competitive marketing quality of the exchanges with these stakeholders.

Regarding the characterization of competitive quality perceptions, it is necessary to:

– obtain information released during these tests which describe the stakeholders' competitive quality perceptions;

– use this information to compare the perceptions produced by the tests to the previous descriptors of the competitive quality perceptions to be generated;

– decide whether any changes need to be made to the process for releasing CQF;

– modify the configurations of CQF emitters in order to carry out these changes;

– characterize the new impacts to obtain on these competitive quality perceptions.

At this stage of the CQP, the quality specialist is fully responsible for the actions to carry out to characterize the stakeholders' competitive quality perceptions.

To fulfill his role, he must participate in all the tasks regarding the way these tests are conducted, who conducts them, and who analyzes the results. It is his role to define how these tasks must take account for the creation of components of the CQP recalled above. He requests the project manager for the authorization to activate these decisions at the correct moment during these stages of the impact studies. He leads the fulfillment of these tasks and evaluates the result of them. These evaluations are presented during some of the project reviews.

The human resources made available are:

– the members of the project team named at time when the impact studies are carried out;

– representatives of the departments in contact with various classes of stakeholders concerned (financial, marketing, purchasing, human resources and those responsible for relations with social regulators).

In terms of possible benefits for the company, these tests can be used to:

– obtain information which describes the signals of an emerging risk of not receiving the compensation expected from the stakeholders;

– estimate the criticality of this risk;

– carry out a diagnosis of causes to better understand the situation, where necessary;

– define corrective action necessary to reduce the criticality of this risk to acceptable levels.

In regards to the use of these tests for validating the expected benefits for the company, the quality specialist simply assists the marketing department, which has full responsibility over these actions.

The use of tests to give a general characterization of the CQF for the future products or services incorporating the innovation falls entirely within the quality specialist's remit.

In regards to the quality of the products or services proposed to stakeholders in the production of the products or services incorporating the innovation, the tests help to:

– supplement the design of the CQT;

– qualif the emitters of CQF integrated into these products or services;

– validate the marketing quality;

– start to envisage methods for supervising the ability of these emitters;

– start to envisage methods for supervising the perceived quality;

– verify the validity of contributions expected from these parties for the production of products or services incorporating the innovation.

Under the responsibility of the project manager, the quality specialist must encourage and assist the industrial department of the company to carry out these actions.

Finally, in terms of contributions to the design of specifications of the CQF of the products or services which are offered to the stakeholders in the innovation process, and that of their emitters, it is simply a case of using some information provided by these tests as primary information to supplement these specifications.

The quality specialist suggests to the project manager to carry out these supplements to the design.

Generally speaking, the role of the quality specialist mainly involves the identification and correction of critical risks of not obtaining the contributions expected from the stakeholders who will use the future products or service, or who will produce them.

He must:

– suggest that these tests include specific methodologies which promote the completion of this objective;

– interpret the information obtained by these methodological tools by identifying the risks and estimating their criticality;

– envisage, or incite others to envisage the changes to implement to reduce the criticality of these risks to acceptable levels;

– suggest the implementation of these corrective preventative actions;

– assist or participate in carrying out these changes;

– compare the results of the change to the objectives;

– estimate the probable effects of this change;

– present an evaluation of the work carried out;

– record the key events.

12.6. The missions of the quality specialist in the feasibility studies

It was shown previously that it is an original CQP that is being implemented, and one that operates in a unique way, throughout the innovation process. It can be repeated that the CQP integrated into the innovation process is also a "one-shot" process. It is therefore essential, like for the innovation, to carry out studies to test its own feasibility.

As before, these studies should take place as naturally as possible in the various stages of the feasibility studies for the innovation.

The project manager is responsible for the correct execution of these studies. He delegates them to the quality specialist because it is important to have specific skills to carry them out, and the other members of the project team will not be asked to collectively acquire these skills on top of their own technical specialties.

The quality specialist cannot carry out these studies alone. As before, he relies on the members of the project team, and the previously mentioned specialized departments, which make their contribution without needing to adopt the particular methodology used.

To give more detail on the actions of the quality specialist, the different stages of the feasibility studies to be reviewed are summarized in the box below.

	CQT	Marketing quality	CQF specifications	Specifications of emitters	Qualification of emitters	Validation of marketing	Supervision of capability	Treating non-compliance	Supervision of perceived quality	Ensuring benefits	Records of improvements
Conducting initial feasibility tests on the future products or services	I	C I	C I	C I							C I
Identifying future stakeholders in the innovation project	I	I							I	I	I
Conducting studies on the nature and feasibility of exchanges with future stakeholders in the project necessary for it to be fully realized	I	I	I	I	I	I	I	I	I	I	I

Conducting performance tests on future products or services	I P	C P	C I P	C I P				I P	I P	C I	C I P
Identifying stakeholders who will contribute to the development and production of future products or services incorporating the innovation	P	P							P	P	P
Conducting studies on the ability of the company to design production and distribution systems for the products and services which will incorporate the innovation by taking into account the configurations of the various definitive drafts of the innovation	P	C P	P	P	C P	C P	C P	C P	P	I P	C I P

Box 12.12. *The CQP in the feasibility studies*

COMMENT ON BOX 12.12.–

Class (C) of stakeholders includes:

– the clients of future products or services incorporating the innovation (the innovation's potential market).

Class (I) includes:

– the stakeholders who produce the innovation (the project team);

– the stakeholders, outside the project team, who provide certain components for the innovation, or resources, which are then transformed by the innovation team to produce the innovation (project providers);

– the stakeholders who invest in the project hoping to reap the benefits from the economic utilization of the innovation (project investors).

Class (P) includes:

– the stakeholders in the BU who produce certain components for the future products or services which will incorporate the innovation;

– the stakeholders outside the BU who will provide certain components of these products or services, or resources which will then be transformed by the company to produce these products or services (providers);

– the stakeholders who will invest in the BU hoping to reap the benefits of producing these products or services (investors).

To provide more detail on these missions, the various stages of the feasibility studies identified in Box 12.12 will be individually reviewed.

12.6.1. *The role of the quality specialist in conducting the initial feasibility tests of the future products or services*

These initial tests take place very early in the innovation process. In general, the project team has not yet been assembled, and the quality department has not necessarily assigned a quality specialist to the project. It is not yet clear whether the idea is worth pursuing.

The various components of the CQP involved in this stage are summarized in the following box.

	CQT	Marketing quality	CQF specifications	Specifications of emitters	Qualification of emitters	Validation of marketing	Supervision of capability	Treating non-compliance	Supervision of perceived quality	Ensuring benefits	Records of improvements
Conducting initial feasibility tests on the future products or services	I	C I	C I	C I							C I

Box 12.13. *The CQP during the stages of the initial feasibility tests of the products or services*

The components of the CQP involved in this stage of the feasibility studies regard two classes of stakeholder: the future users of the products or services incorporating the innovation, and the future contributors to the project.

If the quality specialist has already been named to work on the innovation project, he encourages the project manager to:

– lay the foundations of the CQT to implement during relations with future contributors to the project;

– start to envisage some characteristics of the scenarios of the future processes for the production and use of quality in relations with these two classes of stakeholder;

– start to identify primary information which could be used to design the CQF of the products or services which the project could offer to these two classes of stakeholders and their emitters;

– track these initial ideas.

In general, the creators of the idea naturally feel it necessary to start these reflections. The goal is to see how to attract and motivate the future parties who would allow the project to meet its objectives, and to start to study how to attract future beneficiaries to the innovation.

12.6.2. *The role of the quality specialist in identifying future stakeholders in the project*

The only class of stakeholders affected by the components of the CQP involved in this stage of feasibility studies are the future stakeholders in the project.

The various components of the CQP involved in this stage are summarized in the following box.

	CQT	Marketing quality	CQF specifications	Specifications of emitters	Qualification of emitters	Validation of marketing	Supervision of capability	Treating non-compliance	Supervision of perceived quality	Ensuring benefits	Records of improvements
Identifying future stakeholders in the innovation project	I	I							I	I	I

Box 12.14. *The CQP during the stages of identifying future stakeholders in the project*

The identification of these partners in the project must involve an initial reflection on the CQT which should be drawn up in order for the project to benefit as best as possible from the quality support in exchanges with these socioeconomic partners, and on the marketing quality to be included in the process for the production and use of quality.

This identification must also help to start an analysis of how the quality perceived by these partners could be overseen and how the project could use what these partners have to offer.

The role of the quality specialist is simply to attract the attention of the project manager to the possibility of making a very early contribution to designing certain parts of the CQP, to assist him the researching information which could be included in these parts and to refine them as appropriate.

12.6.3. *The role of the quality specialist in the feasibility studies on exchanges with the future stakeholders in the project*

This stage of feasibility studies is carried out jointly between the project manager and the quality specialist.

The various components of the CQP which are involved in this stage are summarized in the following box.

	CQT	Marketing quality	CQF specifications	Specifications of emitters	Qualification of emitters	Validation of marketing	Supervision of capability	Treating non-compliance	Supervision of perceived quality	Ensuring benefits	Records of improvements
Conducting studies on the nature and feasibility of exchanges with future stakeholders in the project necessary for it to be fully realized	I				I	I	I	I	I	I	I

Box 12.15. *The CQP during the stages of the feasibility studies on the future stakeholders in the project*

The only class of stakeholders is that which includes all the members participating in the production of the innovation.

During these feasibility studies, which can take place in several stages, the quality specialist must ensure that the methods of exchange analyzed integrate the feasibility of the CQT designed in order to reap the maximum benefits from the quality support which will come from exchanges with these partners.

He must validate the scenarios of the processes for the production and use of quality in the exchanges with these partners (competitive marketing quality).

He must qualify the specifications for the CQF of products or services which will be offered as compensation to these partners, and the specifications for the emitters.

He must design and test the feasibility of the procedures: for monitoring the ability of the CQF emitters, for dealing with areas of non-compliance to the CQF specifications, for monitoring the quality perceived by these partners and finally, for evaluating what these partners will supply to the innovation project.

Regarding the feasibility of the CQT, it is above all necessary to:

– verify how the project will produce the products or services aimed at these stakeholders;

– estimate the ability of the stakeholders to produce the services expected by the project in order for it to meet its objectives.

Regarding the qualification of the CQF emitters for the products or services offered to stakeholders, it is above all necessary to:

– identify the risks of not releasing CQF in compliance with the specifications of the emitter, for each CQF emitter;

– draw up corrective actions, where necessary, to undertake in order to reduce the criticality of these risks to acceptable levels;

– validate the implementation of corrective actions;

– write up a qualification report.

Regarding the qualification of marketing quality, it is above all necessary to:

– choose some partners who will be "guinea pigs", for each type of stakeholder. This choice must be taken while limiting the risks of losing an important partner in the innovation process, and ensuring that the test will be representative enough to the situations which will be encountered later;

– in the initial exchanges with these partners, observe how the scenarios develop in terms of producing and using the quality of the products or services that the project offers them;

– formulate any improvements that are necessary;

– validate the implementation of actions carried out by the project manager;

– write up a validation report.

Regarding the constant monitoring of the project to release CQF complaint with the initial specifications, it is above all necessary to:

– ensure that those who activate the CQF emitters verify the ability of these processors to produce the CQF in compliance with their specifications;

– help them, where necessary, to make this verification more efficient.

In terms of monitoring any areas of non-compliance between certain CQF and their specifications, it is above all necessary to:

– ensure that those who activate the CQF emitters send an early alert when an area of non-compliance with CQF specifications is spotted;

– help them to quickly make corrections, if possible, before the area of non-compliance has an impact on the comburant of stakeholders involved;

– encourage stakeholders to quickly alert the project manager if they are victims of an area of non-compliance of CQF;

– ensure that these alerts are followed up by corrective actions.

Regarding the monitoring of stakeholders' competitive quality perceptions, it is above all, necessary to:

– ensure that the project manager regularly surveys stakeholders in order to find out their quality perceptions of the products or services which the project supplies them with;

– help the project manager, where necessary, to correct the CQF of the products or services when quality perceptions do not conform to CQT guidelines.

Finally, in terms of the ensuring the conformity of stakeholders' contributions to the project, it is above all necessary to:

– help the project manager to diagnose the causes of emerging signals of some stakeholders not respecting their commitments;

– at his request, carry out the corrective action required.

This task falls to the quality specialist because they are extremely technical which requires his specific skills to be put to use.

12.6.4. *The role of the quality specialist in carrying out performance tests of future products or services incorporating the innovation*

These tests are mainly technical in nature. They help to control the risks of designing products or services whose specifications do not comply with previous definitions. They also help to reveal possible areas of improvement in their performance. They may be imposed by some future users of the products or services incorporating the innovation (e.g. clinical tests in medicine).

The various components of the CQP which are involved in this stage are summarized in the following box.

	CQT	Marketing quality	CQF specifications	Specifications of emitters	Qualification of emitters	Validation of marketing	Supervision of capability	Treating non-compliance	Supervision of perceived quality	Ensuring benefits	Records of improvements
Conducting performance tests on future products or services	C I P		I	I	C					I	C I P

Box 12.16. *The CQP during the stages of performance tests of future products or services incorporating the innovation*

The main contributions of these tests to the CQP regard the qualification of the CQF emitters aimed at future users of the products or services. This qualification can lead to a modification of the CQF as well as their emitters.

These tests also help to rethink or improve the CQT in exchanges with three classes of stakeholders:

– future users of these products or services;

– partners who contribute to carrying out an innovation process;

– future stakeholders in the BU which will produce the products or services.

These tests can also highlight the information which modifies the CQF, or their emitters, aimed at the partners contributing to carrying out the innovation process, future stakeholders in the BU which will produce these products or services, and certain decision makers.

Regarding the qualification of CQF emitters aimed at future users of the products or services, the actions to carry out can be described as follows:

– simulating the activation of each emitter using the most representative source of information possible from which the primary information will later be drawn;

– observing how the treatment of this information is carried out by the emitter and compare this to the predefined specifications;

– observing the compliance of the information released with the CQF specification;

– identifying the risks of ineffectiveness in this CQF production and estimating their criticality levels;

– formulating corrective actions, where necessary, to undertake in order to reduce the criticality of these risks to acceptable levels;

– validating the implementation of these corrective actions;

– writing up a qualification report.

The quality specialist sets guidelines to ensure that these tests include the actions for the qualification of these CQF emitters aimed at the future users of products or services. Where necessary, he participates in carrying out these tests, ensuring that they comply with the preestablished protocol. He validates the relevance of the results obtained.

He encourages the project manager to carry out modifications, where necessary, to:

– the CQT, to be adopted in exchanges with certain stakeholders on the use of products or services which incorporate the innovation, and on the innovation project;

– marketing quality associated with these stakeholders;

– specifications for CQF and their emitters.

The quality specialist must use the information from these tests to review or supplement the specification of the CQF or their emitters aimed at partners contributing to the implementation of the innovation project and at future stakeholders in the BU which will produce these products or services.

12.6.5. *The role of the quality specialist identifying stakeholders in the future BU*

This stage of the feasibility studies is one which involves the CQP the least. Obviously, it is essential to plan how to make future partners participating in the production of products and services incorporating the innovation adhere to the process as best as possible so that they contribute to its success, but this projection into the future in terms of defining CQP specifications can be greatly supplemented later when it is decided that these new products or services will be designed.

The various components of the CQP which are involved in this stage are summarized in the following box.

	CQT	Marketing quality	CQF specifications	Specifications of emitters	Qualification of emitters	Validation of marketing	Supervision of capability	Treating non-compliance	Supervision of perceived quality	Ensuring benefits	Records of improvements
Identifying stakeholders who will contribute to the development and production of future products or services incorporating the innovation	P								P	P	

Box 12.17. *The CQP during stages of identification of stakeholders in the future BU*

The main stakeholders whom the CQP involves here are:

– the future producers of the production system for the products or services participating in the project;

– the bodies which will finance the design of the production system for the products or services.

The quality specialist does not have an active role in this stage of the feasibility studies. He simply analyzes the decisions which are taken in order to:

– help the project manager supplement the CQT aimed at future stakeholders in the BU which will produce the products or services;

– supplement the methods of estimating the quality perceived by this class of stakeholders;

– supplement the definition of contributions which will be requested from partners in this class.

He must also help the project manager to supplement the CQT regarding certain decision makers, and adapt all other components of the CQP which involve this class of stakeholder.

12.6.6. *The role of the quality specialist in studies on the ability of the company to produce future products or services*

These studies are projections into the future regarding the risks of not being able to produce the products or services which will incorporate the innovation in compliance with their characteristic specifications. Some pieces of information, collected during these studies which can be carried out in several stages, are responses to the demands of future clients.

When the innovation is technical in nature, the future clients can demand, for example, proof showing the ability of the company to master the new technology involved in the innovation.

The various components of the CQP which are involved in this stage are summarized in the following box.

	CQT	Marketing quality	CQF specifications	Specifications of emitters	Qualification of emitters	Validation of marketing	Supervision of capability	Treating non-compliance	Supervision of perceived quality	Ensuring benefits	Records of improvements
Conducting studies on the ability of the company to design production and distribution systems for the products and services which will incorporate the innovation by taking into account the configurations of the various definitive drafts of the innovation	P		I P	I P	P	P	I P	I P	I P	I P	I P

Box 12.18. *The CQP during the stages of studies on the company's ability to produce the future products or services*

Although these studies remain very partial, they produce a reasonable amount of the CQF aimed at several classes of stakeholders:

– the partners, who are likely to speculate on the use of the products or services incorporating the innovation. These may include credit organizations which provide financial assistance to other partners who will use these products or services in order to develop business activities or for personal use;

– the bodies which finance the project;

– the future producers of the production system for the products or services which incorporate the innovation, who are contributing in the project;

– members of the departments within the company responsible for the protection of the environment, people and property.

The information from these studies can lead the project manager to modify various components of the CQP such as the design of the CQT aimed at future producers of the BU, and members of the departments in the company responsible for the protection of the environment, people and property.

They also help to supplement some other components of the CQP which also regard these stakeholders, such as:

– the qualification of CQF emitters;

– the validation of competitive marketing quality.

Finally, they can be used by all stakeholders mentioned above to supplement:

– the monitoring of the capability of the CQF emitters during the release of information;

– the treatment of some non-compliant CQF;

– the monitoring of quality perceived by stakeholders;

– the process of ensuring that the expected compensatory resources will be obtained.

The role of the quality specialist is to:

– suggest to the project manager to finalize the CQT to be adopted in exchanges with future producers of the production system for the products or services incorporating the innovation, and the members of the departments within the company responsible for the protection of: the environment, people and property.

– specify the set of specification for the compensatory resources required from the stakeholders.

He is responsible for:

– overseeing the performance of CQF emitters;

– finalizing the design of the methodology: for monitoring quality perceived by these partners of the products or services offered to them;

– carrying out the final adjustments to the possible contributions of these partners to the resources and the reputation of the company through the innovation.

Of course, as before, the quality specialist is responsible for logging the most significant information highlighted during this stage during these studies which may be used later on.

12.7. The role of the quality specialist in equipment design

The innovation process generally includes stages of equipment design, in order to carry out the process itself, to anticipate the production of products and services which will incorporate the innovation, and to prevent the competition from copying the innovation during the period when this allows the company to occupy shares of the market.

The various stages of the CQP which are involved in the design of production drafts are summarized in the box below.

	CQT	Marketing quality	CQF specifications	Specifications of emitters	Qualification of emitters	Validation of marketing	Supervision of capability	Treating non-compliance	Supervision of perceived quality	Ensuring benefits	Records of improvements
Building a processor responsible for carrying out the innovation process and an external communication network	I	I	I	I	I	I	I	I	I	I	I
Designing prototypes of the production and distribution systems for the products and services	P	I P	I P	I P			I P		I P	I P	I P
Examining the steps to take to ensure the intellectual protection of the innovation			P	P							P

Box 12.19. *The CQP in the stage of equipment design*

COMMENT ON BOX 12.19.–

Class (C) of stakeholders includes:

– the clients of future products or services incorporating the innovation (the innovation's potential market).

Class (I) includes:

– the stakeholders who produce the innovation (the project team)

– the stakeholders, outside the project team, who provide certain components for the innovation, or resources, which are then transformed by the innovation team to produce the innovation (project providers);

– the stakeholders who invest in the project hoping to reap the benefits from the economic utilization of the innovation (project investors);

Class (P) includes:

– the stakeholders in the BU who produce certain components for the future products or services which will incorporate the innovation;

– the stakeholders outside the BU who will provide certain components of these products or services, or resources which will then be transformed by the company to produce these products or services (providers);

– the stakeholders who will invest in the BU hoping to reap the benefits of producing these products or services (investors).

To provide more detail on these missions, the various stages of the equipment design identified in Box 12.19 will be individually reviewed.

12.7.1. *The role of the quality specialist in the construction of the innovation processor*

Building the innovation's production system is the responsibility of the project manager.

It should be recalled that the configuration of a processor is the combination of its resource structure, which may be human, material, or information-based, but also organizational, financial and temporal resources, with its functional potential, i.e. the sum of all its functionalities.

The various components of the CQP which are involved in this stage are summarized in the box below.

	CQT	Marketing quality	CQF specifications	Specifications of emitters	Qualification of emitters	Validation of marketing	Supervision of capability	Treating non-compliance	Supervision of perceived quality	Ensuring benefits	Records of improvements
Building a processor responsible for carrying out the innovation process and an external communication network	I	I	I	I	I	I	I	I	I	I	I

Box 12.20. *The CQP during the innovation processor construction stage*

The stakeholders who may be involved are the following:

– the bodies financing the project;

– future users of the products or services incorporating the innovation, or future producers of the production system for the products or service incorporating the innovation, and who are participating in the project;

– partners outside the project who supply human, material or information-based resources;

– members of the project team.

All components of the CQP can be activated during this stage of the innovation process. They are not set in stone at the end of this stage. As demonstrated before, they can evolve at any moment depending on information collected in certain other stages of the process.

The role of the quality specialist can be summarized as follows:

– suggesting to the project manager to supplement, or review the CQT for this class of stakeholders;

– completing:

- the definition of the marketing quality of these various types of stakeholders, and validating them;

- the design of the CQF for the products or services which are offered to them, and their associated emitters;

- the qualification of these emitters;

- the monitoring of the risks of not releasing CQF in compliance with its specifications;

- the configuration of the treatment system for areas of non-compliance and monitoring quality perceived by these stakeholders regarding the products or services offered to them;

- the method of evaluating the compliance of the contribution to the project with the requirements made by the project manager.

These missions are only carried out through direct collaboration between the quality specialist and the project manager.

12.7.2. *The role of the quality specialist in the design of prototypes of the production system for future products or services*

The innovation process must plan the design of prototypes of the future production system for products or services incorporating the innovation. These concise prototypes are simply intended to control part of the risks of the innovation failing.

The various components of the CQP which are involved in this stage are summarized in the following box.

	CQT	Marketing quality	CQF specifications	Specifications of emitters	Qualification of emitters	Validation of marketing	Supervision of capability	Treating non-compliance	Supervision of perceived quality	Ensuring benefits	Records of improvements
Designing prototypes of the production and distribution systems for the products and services	P	I P	I P	I P			I P			I P	I P

Box 12.21. *The CQP during the stage of designing prototypes of the production system for future products or services*

The objective of these prototypes is to ensure that the company is able to gather the resources necessary to then produce the products or services incorporating the innovation.

The stakeholders affected by the components of the CQP involved in this stage of the innovation process are mainly:

– the investors in the project;

– the future producers of the BU;

– the future investors on the BU;

The actions of the CQP involved are:

– finalizing the CQT regarding future producers and investors in the BU;

– supplementing competitive marketing quality scenarios of exchanges with these partners and the investors in the project;

– producing supplementary information for the CQF of the products or services offered to these three classes of stakeholders;

– monitoring the ability of CQF emitters to comply with their objectives during this stage and for all three classes of stakeholders;

– carrying out the final adjustments to the possible contributions of these partners to the resources and the reputation of the company due to the innovation.

The role of the quality specialist is to suggest to the project manager to:

– finalize the CQT;

– accurately identify the specifications for the compensatory resources requested from them.

He is responsible for monitoring the production of supplementary information for the CQF, and monitoring the capability of the CQF emitters.

12.7.3. *The role of the quality specialist in the intellectual protection of the innovation*

This protection responds to a need for all partners who contribute in the production of the innovation process or who will be the targeted clients of the future products or services which incorporate the innovation.

This stage of the innovation process therefore plays a role in the CQP associated with it.

The components involved are the following.

	CQT	Marketing quality	CQF specifications	Specifications of emitters	Qualification of emitters	Validation of marketing	Supervision of capability	Treating non-compliance	Supervision of perceived quality	Ensuring benefits	Records of improvements
Examining the steps to take to ensure the intellectual protection of the innovation			I P	I P							

Box 12.22. *The CQP during the stage of intellectual protection of the innovation*

The components of the CQP which are only stages of design for the CQF and their emitters are carried out naturally without the intervention of the quality specialist.

12.8. The role of the quality specialist in assisting decision-making

The decision makers are key stakeholders in an innovation process. In our model, the CQP which involves them is integrated into two stages of the innovation process which regard assistance in decision-making. They are products or services offered by the project to decision makers to subsequently obtain decisions regarding whether to continue with the innovation process, and on the ways of continuing. They are intended to help to control the risks of their decisions and grasp the opportunities regarding an optimal economic use of the innovation.

The CQP integrated into the innovation process is intended to use the quality perceived by the decision makers – who use this assistance to make their decisions – in order to prompt decision which lead to the success of the innovation.

Care should be taken here; however, these decisions are not necessarily compliant with the expectations of the project team members. It is not a question of using quality to influence the positive decisions being taken by these partners, but of helping these decision makers take the decisions which will, in the end, be the most beneficial for the success of the innovation. An insightful decision may be at odds with what these partners wish, since they only have a partial vision of the company's socioeconomic environment. Consequently, for the project, the point of a decision is for it to have a positive impact on the success of the innovation.

All the components of the CQP which are involved in these products or services can be summarized in the following box.

	CQT	Marketing quality	CQF specifications	Specifications of emitters	Qualification of emitters	Validation of marketing	Supervision of capability	Treating non-compliance	Supervision of perceived quality	Ensuring benefits	Records of improvements
Creating a rationale to aid decision-making, examining the probable socioeconomic impacts of implementing the innovation	D	D	D	D	D	D	D	D	D	D	D
Creating a rationale to aid decision-making which details the feasibility of implementing the innovation	D	D	D	D	D	D	D	D	D	D	D

Box 12.23. *The CQP during the decision-making stages*

The decision makers affected by the creation of a rationale to assist decision-making, which examines the probable socioeconomic aspects of using the innovation, are:

– investors in the company;

– the commercial department;

– the management of the BU which will produce the products or services incorporating the innovation;

– the “R&D” department;

– those responsible for sustainable development.

The decision makers affected by the creation of a rationale to assist in decision-making which examines the feasibility of using the innovation are:

– investors in the company;

– the management of the BU which will produce the products or services incorporating the innovation;

– the purchasing department;

– the “R&D” department;

– the industrial department

The actions to carry out in order to design the CQT can be described in the following ways:

– describing the macro-characteristics of the demands, or expectations, or decision makers regarding the assistance offered to them for which it is necessary to attempt to satisfy, in order to benefit from constructive decisions;

– identifying the CQT of the main competitors which could have an effect on decision-making;

– identifying the expectations, which are not being satisfied by competitors, to which decision makers attach great importance, and to which the project is able to better satisfy;

– understanding the behavioral mechanisms of the decision makers, which forms the basis of their quality perceptions of this assistance, leading them to express a competitive quality judgment (these mechanisms will be subsequently referred to as "quality comburant");

– describing and creating a hierarchy of these quality perceptions according to the significance of their impact of the competitive quality judgment;

– describing the main characteristics of the desired decisions from most decision makers in exchange for assistance in making decisions (expected benefits);

– estimating the capacity of decision makers to produce them;

– describing the uses which will be made up of them by the project team;

– estimating the interest perceived by decision makers to provide the effort required in order to use the decision-making assistance offered by the project;

– giving a very general characterization of the CQF of decision-making assistance to provide;

– carrying out a general feasibility study on the release of information which will make up the CQF.

COMMENT 12.1.– This particular example of implementing mechanisms for designing CQT shows the power of such a process, but also all the difficulties of finding useful information and making an insightful judgment on this information.

The role of the quality specialist at this stage of designing the CQT to be adopted in relations with decision makers is the "coaching" of the project manager.

He must observe how the project manager naturally undertakes these tasks and suggest supplementary actions to him, where necessary, or corrections, at the request of the project manager.

The quality specialist must always remain withdrawn in these exchanges with decision makers.

The actions to carry out to design these scenarios of the role-plays between the members of the project team and the decision makers, which will allow that the quality of the exchanges to contribute to the success of the innovation can be described in the following ways:

– identify the events which are:

- likely to be produced by the project, and the decision-making assistance, possibly coproduced with these stakeholders, which form the basis of the information which will stimulate the decision makers' competitive quality perceptions by complying with CQT provisions;

- likely to be observed, which describe how information in the CQF of decision-making assistance will have an impact on the decision makers competitive quality perceptions;

- likely to be observed, which will allow the members of the project team to make a judgment on the contributions of the decision makers to the success of the innovation process;

– describe a predefined and organized flow diagram of each of these events;

– present this diagram to some decision makers in order to obtain their reactions;

– interpret their reaction in terms of opportunities or risks regarding the effects of their competitive quality perceptions on the insightfulness of the decisions taken;

– adapt the initial diagram to better comply with the provisions of the CQT.

COMMENT 12.2.– With this example, it can be noted that what is known as competitive marketing quality is a collection of actions which are often undertaken in a natural way during the project.

The quality specialist is the real director for these role-play scenarios between the members of the project team and the decision makers, who will allow the quality of the exchanges to contribute the success of the innovation. However, these scenarios are only submitted to the project manager if he so requests. If the project manager does not wish to be assisted in this task, this can be used as guidelines to help the quality specialist to fulfill other objectives.

The aspects of design of the CQF which must be integrated into the decision-making assistance process are:

– the reminder of the partial descriptions of the decision makers' competitive quality perceptions to prompt the CQF, which are included in the CQT;

– the identification of primary information taken advantage of during previous stages of the innovation process which can be used to design the CQF;

– planning the use of this information in marketing quality scenarios for exchanges with decision makers;

– the identification of vehicle through which to release the CQF by certain subprocessors of the processor which is carrying out the innovation process, some events of the innovation process, the components of the innovation's configuration and the effects to which it is likely to cause on its environment;

– the design of the specifications of the CQF to be released to prompt competitive quality perceptions which comply with CQT provisions;

– the design of scenarios of the process of transforming primary information into quality fuel which complies with the specifications;

– the designs of the processors, which will carry out the collection of primary information, transform it into CQF and communicate this CQF to decision makers.

The validation of marketing quality to specifications of CQF and the qualification of emitters is carried out from the first exchange with decision makers.

In terms of the design of the CQF which must be included in the decision-making assistance process, the quality specialist carries out a technical task of regulation which consists of:

– observing how the project manager designs the content and form the decision-making assistance;

– extracting information from these observations which describe the CQF and their emitters;

– offering the project manager a preanalysis of the risks of not prompting the competitive quality perceptions desired by decision makers;

– completing this analysis, with the project manager, by highlighting any possible risks of decisions likely to reduce the chances of the innovation succeeding, with an unacceptable criticality level;

– suggesting a search for solutions to reduce the criticality to acceptable levels;

– assisting, or participating in, this quest for solutions, especially taking advantage of feedback as a guide;

– estimating the probable effects of the changes undertaken compared with the objectives;

– recording the key events of these improvements.

Monitoring the aptitude of the project manager to produce this decision-making assistance is a self-monitoring process.

This consists of ensuring that the CQF and their emitters are closely compliant with the predefined specifications, just before the production of the decision-making assistance.

The quality specialist intervenes at this stage only if the project manager requests him to do so.

Nevertheless, the quality specialist must monitor how the information released by the decision-making assistance impacts on their competitive quality perceptions. He can only do this by interpreting the signals of the decision makers' attitude and behavior.

Subsequently, the quality specialist carries out a critical analysis of the deviations between the competitive quality perceptions prompted and those which were expected, and the reasons for these deviations. He presents this diagnosis to the project manager. Such deviations do not always have negative effects. They can highlight opportunities to capitalize in order to increase the chances of the innovation succeeding.

This diagnosis can lead to the implementation of corrective action which takes the form of supplementary decision-making assistance.

12.9. Conclusion

In this chapter, we have tried to offer guidelines for the work of the quality specialist in the innovation process. These guidelines simply represent what we have learnt from feedback on experiences.

Each company and each quality department involved in an innovation process can use these guidelines and adapt them to the specific situations encountered.

It would be desirable for the role of the quality specialist to be formally described in one of the annexes of the company's quality manual.

This would help to strengthen the confidence of various parties in the innovation process.

The company must also allow the quality department, responsible for the quality specialist working on the innovation project, to carry out internal audits.

As much as the confidentiality necessary in each innovation process limits the specific practices of each company to produce and use the quality of the innovation, it is just as necessary to promote a benchmark for the quality specialists' practices in these processes.

13

Quality Culture in Project Teams

13.1. Introduction

All the partners in the project team involved in an innovation process contribute to the efficiency of the competitive quality process (CQP) process. It has been shown that the quality specialist can take the role of "coach" and must directly carry out certain technical actions so as not to overload these partners who use specific methodological skills in the CQP.

These partners' contributions are limited to the understanding of the usefulness and logic of the various components in the CQP and the production of relevant information which can help to carry them out. They can either independently ensure this production, or act as one of the partners in the process for the production of information led by the quality specialist.

For these contributions to be optimal, it is necessary for them to have, or acquire, a certain culture known as "a quality culture specific to the innovation process" (QCI).

The aim of this chapter is to specify the guidelines for this culture.

13.2. The inherent visions of quality

The first component of a QCI is the brief and predefined vision that the individual has of quality. This vision depends on his personality, and the influence from his social environment.

Sociological studies carried out by IRDQ Sociological studies show that this vision is a combination of basic visions which can be described in the following way:

– *artisanal vision*: the individual believes that producing quality means feeling that something has been "well-made";

– *engineering vision*: the individual believes that producing quality means mastering the technology integrated into the products;

– *regulatory vision*: the individual believes that producing quality means conforming to standards;

– *marketing vision*: the individual believes that producing quality means satisfying the clients;

– *egocentric vision*: the individual believes that producing quality means having an enjoyable experience in the exchange with the other party;

– *collectivist vision:* the individual believes that producing quality means meeting the objectives set by the group.

Each individual has in mind these various basic visions, known as "inherent visions of quality", and they rank them in a personal way.

The creator tends to focus their vision of quality on a combination of the "artisanal" and "egocentric" visions. The inventor tends to focus their vision of quality on a combination of the "artisanal", "engineering" and "egocentric" visions. Some quality specialists tend to focus their vision of quality on a combination of "regulatory" and "marketing" visions. A director tends to focus their vision of quality on a combination of "collectivist" and "marketing" visions, and so on.

These examples show that it is difficult to make members of a group stick to a shared quality project because they have visions of quality which are intrinsically different.

It is particularly difficult in an innovation process because the members of a project team have assertive personalities. It is difficult during a project to try to converge their quality visions toward one common vision. This is often a task undertaken by quality specialists integrated into the innovation project. We believe this to be a mistake. Our experience tends to show that it is actually necessary to use the intrinsic variety of these quality visions which the members of the project have in order to give the maximum added value to the final production of quality in the innovation. It is up to the quality specialist to assist the project manager in using this variety for the collective good. An innovation process must above all be the emergence of various ideas which constitute the energy of the innovation process. The energy of the integrated CQP, therefore, also relies on the variety of the visions of quality held by the members of the project team.

The quality specialist must have a vision of quality with the widest scope. They must be able to build an overview of the quality visions of all the members of the project team. They must see themself as the conductor of an orchestra who is synchronizing the quality visions of all the members of the project team.

13.3. The aptitudes for producing quality

Two classes of aptitudes can be distinguished:

– individual aptitude to contribute to the production of quality in the innovation process (IAQ);

– collective aptitude to contribute to the production of quality in the innovation process (CAQ).

IAQ is a collection of attitudes and behavior which are produced when one person carries out an action on a CQP component.

CAQ is a collection of attitudes and behavior which are produced when the person prompts other people to carry out a component of the CQP.

13.3.1. *Individual aptitudes*

In general, members of an innovation process are asked to adopt three types of aptitude:

– aptitude to carry out "standard" activities;

– aptitude to carry out activities with "varieties at the source";

– aptitude to carry out personalized activities ("residual varieties").

An aptitude for carrying out "standard" activities is an aptitude to systematically reproduce the same attitude or behavior by complying with models.

An aptitude to carry out "variety at the source"-type activities is an aptitude to spot various pre-established situations, and to systematically adopt for each of them, adapted "standard" attitudes and behavior.

An aptitude to carry out "residual varieties"-type activities is an aptitude to adapt as best as possible attitudes and behavior to each situation encountered while taking their own experience into account.

A worker must have these three types of aptitudes and use them appropriately. This is the difference between a robot and a human being. One of the great dangers of implementing a competitive quality tactic (CQT) can come from the initiative of some workers who, in order to perform better or be better noticed, adopt personalized attitudes or behavior which harm the quality perceived by some stakeholders.

A classic example is that of a waiter in a restaurant who disturbs you every five min to check that everything is fine.

In an innovation process, "standard"-type attitudes are often reserved for a production of information which supplies the components of the competitive quality fuel (CQF) which influences stakeholders' perceptions of trust, by satisfying security demands. These attitudes are the most restrictive. They are concentrated in the feasibility study stages.

The "varieties at the source"-type attitudes are most often requested to apply common organizational rules for collective production within the innovation process. Compliance with some of these rules can have an influence on the performance of some CQF components. These are most often encountered in impact studies where uncertainties are often linked to stakeholders' attitudes and behavior.

"Residual varieties"-type attitudes are generally those which contribute the most to the production of the added value of the innovation. They are at the heart of the innovation process' creative research. They play a key role in the production of information which makes up the CQF.

A member's IAQ in the implementation of these aptitudes for the benefit of the CQP can be expressed by characteristics which can be described in the following way:

– personal interest to contribute to carrying out a CQP component;

– awareness of the personal role which is played by this contribution;

– individual behavior in the self-regulation of actions which produce this contribution;

– personal attitudes in relations with other contributors to the production of this CQP component.

Each of these characteristics can be recognized by specific signals which can be summarized in the following way.

The signals which express the personal interest to contribute to carrying out a CQP component are:

– how the means of enjoyment are found while contributing to produce the CQP component;

– how to be recognized by this contribution;

– how to use this recognition in order to obtain the compensatory resources expected.

The signals which express the awareness of the personal role which is played in this contribution are:

– how contributions to the objectives of the CQP component are described;

– how these contributions are located in relation to those of other members of the project team;

– how to identify the needs of stakeholders which benefit from the results of these contributions;

– how to characterize the performance necessary for the configuration of their workspace in order to make the most effective contribution to the CQP component;

– how to characterize the performance to be required of the resources which will supply their workstation in order to provide the most effective contribution.

The signals which express individual behavior in self-regulation of the actions which produce this contribution are:

– self-monitoring of actions in order to avoid failure in methods;

– reactions which avoid, or compensate for, the effects of failure in methods;

– elimination of causes which influence these methodological failures.

Signals of personal attitudes in relations with other contributors in the production of this CQP component are:

– "active listening" of the information, given by partners, which expresses requirements or its competitive quality perceptions of the products or services which are provided to them;

– "reactive listening" of the claims, or supplementary requirements, given by the partners, when the responses can improve the competitive quality perceptions of the products or services which are provided;

– the expression of expectations in terms of "receivables" given by the partners, regarding the resources which are provided to them;

– the description of characteristics of the products or services provided to them to prove their quality, in such a way so as to enhance them and make them accessible;

– the collaboration with partners to co-produce the CQP component.

Active and reactive listening systematically include the actions of:

– a precise grasp of the information released by the partner in question;

– an evaluation of the deviations between what is expressed by the partner and what had been predicted by the project beforehand;

– the decision regarding the choice of responses to deal with these deviations;

– the implementation of decisions for coproduction with the partner in question.

13.3.2. *Collective aptitudes*

CAQ is expressed when an individual leads a group of workers in carrying out a CQP component.

In the case of an innovation process, usually it is the project manager or the quality specialist who is in charge of leading such groups. However, other members of the project team may also make up small groups to combine actions, give an opinion on a situation and take decisions. The complexity of situations, encountered during the CQP integrated into an innovation process, means that it is necessary to gather all the project team members' partial, but complementary visions.

It is possible to detect the collective attitude of a partner to contribute to a CQP component in the following way:

– it is presented in a comprehensible language which specifies the objective to be met;

– it gives rise to inherent visions of quality of each participant by requesting them to participate in the validation of the objective;

– it uses inherent visions to characterize the importance of the contribution of the objective to the CQP's efficiency;

– it distributes the scenario of the process for the group's production;

– it provides a dynamic regulation of the process;

– it distributes the responsibility of the results obtained between the participants.

Dynamic regulation consists of:

– having the group validate the evolutions of the objectives and the direction to following order to meet them;

– constantly keeping the criticality of pre-identified risks of not meeting these objectives at levels which are acceptable for the group;

– correcting the deviations from the guidelines, considered by the group as risks with unacceptable criticality levels;

– having the group validate the evaluation of the shared effects of the results obtained on the effectiveness of the CQP component;

– carrying out, where necessary, any corrective action to improve the impact of the results on this effectiveness.

All these culture-based guidelines are discussed further in our book entitled "*Best Practices in TQM*".

It can be observed that in those companies which use quality to strengthen their competitiveness through innovation, these interests, roles, behavior and attitudes gradually become part and parcel of the individual and collective culture of the company. This culture generally helps to gradually reduce the restrictions and artificial guidelines (procedures) which are necessary in the earlier learning stages. It naturally generates high levels of transparency, and great flexibility in activities and exchanges. However, the mobility of workers means that this quality culture must be passed on by working in partnership and by following very operational training, often based on simulations of real situations.

The 20th Century has given rise to many models and methodologies intended to help companies to control the quality of their products and services. This situation is due to the growth of their complexity, their production levels, and demands of consumers and public authorities which have benefited from competitive pressure which has developed in all sectors of the economy. This proliferation and sophistication of tools available to companies tends to overshadow the production and use of quality in relations between the company and its stakeholders mainly regarding behavior and attitudes of workers which produce added value.

It is very clear that this very analytical modeling of the companies' competitive quality culture that we have shown is simply one method of tracking it. It is important not to ask each team member, either before or during his/her work, to follow these indications to the letter. The principle itself of a culture is that these various cultural factors must be rooted in the instinct which the member employs to carry out a given action. The members gradually acquire these instincts by stages of learning. They will continue to develop as he/she is surrounded by an environment where the other company members have the same instincts (contagion effect).

A culture is of course a form of conditioning which is necessary to carry out collectively in order to reach a goal. This competitive quality culture is a very sensitive asset to a company that continually needs to be maintained and strengthened, and the effects that it produces on the company's general competitiveness are long-term in nature. This is partly why the directors are hesitant about investing heavily in this area and prefer to resort to other means with more rapid effects, in order to develop their short-term competitiveness. Unfortunately, in the medium or long-term, these policies often act against the interests of the company.

13.4. Methods for acquiring aptitudes for producing quality

This is an important piece of work to carry out in research centers, companies, competitiveness clusters and incubators in order to highlight the real significance of quality in ensuring the success of innovation processes.

The messages usually given to the members of these bodies are worrying. They speak of quality processes in terms of management standards, or client satisfaction. We are very far away from the concerns of these members when they make predictions about an uncertain future based above all on their imagination and the rigorous rule of scientific research. What is the value in spreading fear by sending messages which could be perceived as blaming, technocratic or limiting the ambitions of the project to satisfy the distant and debatable expectations of clients. We have taught quality in doctoral training for several years in a pragmatic way, by discussing students' research and development activities and by regularly showing how to use quality concepts step-by-step in order to move their work forward in a more controlled manner.

This experience, associated with advisory services provided to companies, has driven us to advocate an approach focused on quality culture in the innovation process simply through a learning program regularly included in the process at moments when it is clear to see the advantage of using quality.

The quality culture is not something which is imposed. It is uncovered through gradual practice of the workers who acquire it as they are constantly requesting new information regarding useful practices involving quality.

13.5. Conclusion

The effectiveness of a CQP in an innovation process is essentially based on the creative quality culture of the project team members. The quality specialist must be the first to acquire this specific culture. It is, therefore, necessary for them to be able to closely connect the scientific and quality cultures, without looking for any personal enhancement by imposing inappropriate rules to the specifics of research and development. This is why we believe that quality specialists, with two cultures, must have come from the field of research, and acquire training in quality management in the specific sector of the economy in which they work.

Appendices

Appendix 1

Main Guidelines for an Innovation Process

A1.1. Innovation Process

An innovation process consists of the following stages:

– deciding to foster new ideas to help the progress of the company;

– bringing about the emergence of an idea which may provide a response to a need of society;

– briefly characterizing products or services which could benefit from the innovation;

– identifying possible clients for these future products or services;

– characterizing future uses for products or services which would incorporate the innovation;

– creating experimental drafts which represent the idea as accurately as possible;

– conducting initial feasibility tests on the future products or services;

– deciding to initiate prospective studies on how to use the idea to benefit the company;

– conducting prospective studies on possible socioeconomic uses for the innovation, and its impacts on the company and its position;

– deciding to guarantee the ability of the company to mobilize the resources necessary, internally and externally, in order to fully realize its innovation process;

– identifying future stakeholders in the innovation project;

– conducting studies on the nature and feasibility of exchanges with future stakeholders in the project necessary for it to be fully realized;

– deciding contractual issues with these stakeholders, and building a processor to carry out the innovation process;

– building a processor responsible for carrying out the innovation process and an external communication network;

– deciding to progressively design drafts which give a more accurate representation of the idea;

– progressively designing these drafts to accurately represent the configuration of the innovation;

– deciding to carry out prospective studies on the socioeconomic impacts of the future products and services which will incorporate the innovation;

– progressively designing drafts which give a representation of future products and services to incorporate the innovation, by describing in particular, the future "interfaces" of the innovation with other components of these products and services;

– conducting tests on the drafts of future products or services, at various steps in their production, on future clients and on all future stakeholders in the production system for the products and services;

– conducting performance tests on future products or services;

– creating a rationale to aid decision-making, examining the probable socioeconomic impacts of implementing the innovation;

– deciding to guarantee the ability of the company to adopt a production system for the future products and services;

– more precisely, identifying stakeholders who will contribute to the development and production of future products and services which will incorporate the innovation;

– designing drafts of the production and distribution systems for the products and services;

– conducting studies on the ability of the company to design production and distribution systems for the products and services which will incorporate the innovation by taking into account the configurations of the various definitive drafts of the innovation;

– creating a rationale to aid decision-making which details the feasibility of implementing the innovation;

– deciding to validate the innovation;

– planning the steps to take in order to ensure the intellectual protection of the innovation.

A1.2 Creative research

Creative research regards the following stages of the innovation process:

– bringing about the emergence of an idea which may provide a response to a need of society;

– creating experimental drafts which provide a more accurate representation of the idea;

– progressively designing drafts to accurately represent the configuration of the innovation;

– progressively designing drafts which give a representation of future products and services to incorporate the innovation, by describing in particular, the future "interfaces" of the innovation with other components of these products and services.

A1.3. Impact studies

Impact studies regard the following stages of the innovation process:

– briefly characterizing products or services which could benefit from the innovation;

– identifying possible clients for these future products or services;

– characterizing future uses for products or services which would incorporate the innovation;

– conducting prospective studies on possible socioeconomic uses for the innovation, and its impacts on the company and its position;

– identifying stakeholders who will contribute to the development and production of the future products and services which will incorporate the innovation;

– conducting tests on the drafts of future products or services, at various steps in their production, on future clients and on all future stakeholders in the production system for the products and services.

A1.4. Feasibility studies

Feasibility studies regard the following stages:

– conducting initial feasibility tests on the future products or services;

– identifying future stakeholders in the innovation project;

– conducting studies on the nature and feasibility of exchanges with future stakeholders in the project necessary for it to be fully realized;

– conducting performance tests on future products or services;

– conducting studies on the ability of the company to design production and distribution systems for the products and services which will incorporate the innovation by taking into account the configurations of the various definitive drafts of the innovation.

A1.5. Organizational stages

The organizational stages are as follows:

– building a processor responsible for carrying out the innovation process and an external communication network;

– creating a rationale to aid decision-making, examining the probable socioeconomic impacts of implementing the innovation;

– designing prototypes of the production and distribution systems for the products and services;

– creating a rationale to aid decision-making which details the feasibility of implementing the innovation;

– planning the steps to take in order to ensure the intellectual protection of the innovation.

A1.6. Decision-making stages

The decision-making stages are as follows:

– deciding to foster new ideas to help the progress of the company;

– deciding to initiate prospective studies on how to use the idea to benefit the company;

– deciding to guarantee the ability of the company to mobilize the resources necessary, internally and externally, in order to fully realize its innovation process;

– deciding contractual issues with these stakeholders, and building a processor to carry out the innovation process;

– deciding to progressively design drafts which give a more accurate representation of the idea;

– deciding to carry out prospective studies on the socioeconomic impacts of the future products and services which will incorporate the innovation;

– deciding to guarantee the ability of the company to adopt a production system for the future products and services;

– deciding to validate the innovation.

A1.7. Main evaluation guidelines

The following are the main evaluation guidelines for an innovation process:

– the evolution in market uptake in terms of volume;

– the speed and size of returns on investment;

– the evolution in the purchasing value given by the market to the products or services which incorporate the innovation;

– the evolution in the attraction to the innovation compared to that of competitors' products or services;

– the nature and extent to which the innovation provided an answer to society's evolving needs;

– the level of contribution to the enhancement of the fundamental image of the products or services which incorporate the innovation, or to the fundamental image of the company;

– the level of enhancement of the "expertise" and culture of the company;

– the ease of finding and using this new "expertise" necessary for the production of new products and services;

– the time taken to introduce the products or services which incorporate the innovation into the market;

– the contribution of the innovation to the fulfillment of the company policy.

Appendix 2

Guidelines for a Competitive Quality Process

A2.1. The main classes of competitive quality perceptions of an innovation process

– Perceptions of expectancy regarding:

- the satisfaction of expectations by certain components of the innovation production process, its results or the effects it causes;

- the ability to use these components;

- the ability to make the work required to understand, obtain, adjust and use them profitably.

– Perceptions of confidence in the ability of the innovation process to comply with the objectives given.

A2.2. The technical function of a competitive quality process integrated into an innovation process

The technical functions of a competitive quality process (CQP) integrated into an innovation process can be described as follows:

– define a competitive quality strategy in exchanges with stakeholders in the innovation process;

– for each class of stakeholder, design scenarios regarding production and release of competitive quality fuel, competitive quality judgments and contributions from stakeholders which have the best chance of fulfilling the competitive quality strategy (marketing quality);

– design specifications for competitive quality fuel (CQF) offered to stakeholders, so that scenarios have the best chance of being fulfilled with the minimum amount of energy expended (design of competitive quality fuel);

– design emitters of CQF integrated into the production systems of products or services offered to stakeholders, capable of systematically complying with the CQF specifications (design CQF emitters);

– when each CQF emitter is activated, ensure that it is able to systematically comply with these specifications (qualification of CQF emitters);

– check the compliance of the various processes related to marketing quality to their scenario (validation of marketing quality);

– ensure, during the production process of competitive quality fuel, that the risks of non-compliance to their specification maintain an acceptable level of criticality (supervision of the capability of CQF emitters);

– eliminate the areas of non-compliance with specification as soon as possible in the CQF production process (processing of non-compliance with specifications of competitive quality fuel);

– oversee the impacts of CQF in the innovation process on the competitive quality perceptions of various stakeholders (supervision of perceived quality);

– ensure that the stakeholders' competitive quality perceptions lead to the supply of compensatory resources in compliance with the expectation of the competitive quality strategy (ensuring expected benefits);

– track feedback in order to improve the efficiency of competitive quality processes which will be integrated into new innovation processes (improving the ability of the company to use stakeholders' competitive quality judgments on an innovation process).

A2.3. Lines of actions for the various components in the competitive quality process

A2.3.1. *Competitive quality tactics (CQTs)*

The main areas of input data in the process of designing CQT can be described as follows:

– description of macrocharacteristics of demands or expectations of the stakeholder regarding the products or services which have already been offered, or that must be satisfied in order to benefit from its services;

– identification of CQT of the main competitors which may disrupt exchanges with the stakeholder;

– identifying the expectations not being satisfied by the competition, to which the stakeholder attaches a great deal of importance, and that the project is able to satisfy better;

– discovering the stakeholder's behavioral mechanisms which are at the basis of these quality perceptions in terms of satisfying expectations, and which prompt a competitive quality judgment (these mechanisms will then be grouped under the heading "stakeholder's quality comburant");

– describing and establishing a hierarchy of quality perceptions in accordance with the importance of their impacts on its competitive quality judgment;

– describing the main characteristics of the services that are expected to be received from the stakeholder in exchange for the products or services which it is offering (expected benefits);

– estimating the capacity of the stakeholder to produce them;

– describing the uses that the company will make of them;

– estimating the perceived interest from the stakeholder to provide the work requested in order to obtain the products or services offered by the company;

– giving a very general characterization of the "quality fuel" (definition 2.3) of the products or services which will be offered to the stakeholder in compensation for these services;

– carrying out a very general feasibility study on the release of the information which will make up the CQF of the products or services.

A2.3.2. *Competitive marketing quality*

– Identifying events which are:

- likely to be produced by the project, the products or services offered to stakeholders, or coproduced with stakeholders, which release information to stimulate their competitive quality perceptions, while complying with the CQT provisions;

- likely to be observed, describing how the information in the quality fuel of the project's products or services offered to the stakeholders will have an impact on their competitive quality perceptions;

- likely to be observed, helping a judgment to be made on the compliance to the initial specifications of the stakeholders' contributions to the success of the innovation process;

- likely to be observed, helping a judgment to be made on how the stakeholders may influence the image or reputation of the company.

– Describing a predefined and organized flow diagram of each of these events.

– Presenting this diagram to certain stakeholders to gauge their reactions.

– Interpreting their reactions in terms of opportunities or risks with regard to the effects of their competitive quality perceptions, the provision of compensatory resources to the project or future business unit (BU) and their contribution to the company image or reputation.

– Modifying the initial diagram to comply with provisions in the CQT.

A2.3.3. *Design of competitive quality fuel*

– Recalling CQT data regarding the competitive quality perceptions to be prompted for a given class of stakeholders.

– Ascertaining the nature of the information which may be released to prompt these perceptions during various stages of the innovation process.

– Placing the release of this information into the scenarios of competitive quality marketing.

– Locating the various vehicles possible for this information by identifying subprocessors of the main processor carrying out the innovation process, the events of the innovation process, components of the innovation's configuration and the effects that it may produce on its environment which may release this information.

– Designing the specifications of each piece of information so that they have the best chance of prompting these perceptions when released.

– Validating these specifications by using various drafts of the innovation, and future production and distribution systems of the products or services which will incorporate it, predictions on future uses of the products or services, as well as carrying out performance tests on future products or services.

A2.3.4. *Design of CQF emitters*

– Spatial and temporal localization of the five processors identified above, possible productions of primary information which could be used to release future CQF.

– Highlighting the characteristics of this primary information.

– Designing scenarios in the process of transforming this primary information into CQF which complies with their specifications.

– Designing the processor which will obtain the primary information, transform it into CQF and communicate the CQF to stakeholders.

A2.3.5. *Qualification of CQF emitters*

– Prompting the activation of the emitter by using the most representative source of information possible, from which raw data will be drawn.

– Observing the methods of extraction of primary information necessary for the communication of information which complies with the CQF specifications imposed.

– Identifying inefficiency risks in obtaining this information and estimating their criticality.

– Formulating any corrective actions necessary to reduce the criticality of these risks to acceptable levels.

– Validating the implementation of these corrective actions.

– Producing a qualification report.

A2.3.6. *Validation of competitive marketing quality*

– Choosing a stakeholder who will fulfill the role of “guinea pig”. This choice must be taken, limiting the risk of losing an important partner in the innovation process, but ensuring that the test is sufficiently representative of the situations which will be encountered later on.

– Nominating observers based on their ability to spot the critical points in the four processes. These observers must be able to observe the situation from various angles. They must be just as perceptive at identifying and creating a hierarchy of risks, as spotting new opportunities emerge. They must be familiar with the scenarios envisaged for the processes to be observed, and the specifications attributed to the CQF and their emitters.

– Directing observations with guidelines established during the previous stages of the CQP and by appropriate methodological tools;

– Planning this validation at the correct moment in the innovation process.

– Preventing bias being introduced during these observations by insisting the team members involved in the process adopt an attitude that they could not reproduce later on.

– Carrying out a critical analysis linking the facts featured in the overview to probable effects and possible causes. A hierarchy of FMECA-type methods then helps to rank the contribution of facts to the non-compliance with provisions of the initial CQT.

A2.3.7. *Supervising the capability of the production of products or services*

– Verifying, before each sequence of an emitter releasing quality fuel, that its configuration complies with the specifications of its design, and that the source of primary information, that it must obtain and transform, is easily accessible. When a risk of not releasing the quality fuel which complies with these specifications is observed, and the criticality of this risk is at an unacceptable level, corrective action needs to be taken.

– Observing information released by the emitter.

– Comparing it with the specifications of the quality fuel which it must release.

– Deciding how to correct the deviations observed.

– Tracking corrections.

– Tracking the corrective action taken in order to capitalize on this knowledge and expertise.

A2.3.8. *Supervising areas of non-compliance*

Corrections are made either to:

– the quality fuel itself, to make it more compliant with the specifications before it influences the stakeholders' quality comburant;

– the causes regarding the way in which the emitter was activated, or in its configuration;

– the compensation of the effects produced on stakeholders' competitive quality perceptions.

The creation of this ranking guides the project manager's decisions of whether or not to implement an improvement process regarding the performance of the various stages encountered previously in the CQP.

This specific process consists of the following stages:

– obtaining information which shows stakeholders' competitive quality perceptions;

– interpreting this information with regard to the CQTs;

– deciding any direction changes which need to be made in the process of releasing CQF;

– modifying the configurations of their emitters in order to carry out these direction changes;

– characterizing the impacts obtained.

A2.3.9. *Supervision of perceived quality*

– Obtaining the information which shows stakeholders' competitive quality perceptions.

– Interpreting this information with regard to the CQTs.

– Deciding any direction changes which need to be made in the process of releasing CQF.

– Modifying the configurations of their emitters in order to carry out these direction changes.

– Characterizing the impacts obtained.

A2.3.10. *Ensuring benefits*

– Obtaining information which describes signals of emerging risks that stakeholders will not provide the expected compensation.

– Estimating the criticality of this risk.

– if Necessary, carrying out a diagnosis of causes to better understand the situation.

– Defining the corrective action necessary to reduce the criticality of the risk to acceptable levels.

– Carrying out this corrective action.

– Evaluating the impact on the criticality of the risk.

Appendix 3

The Role of the Quality Specialist in an Innovation Process

Stages in the innovation process	**Nature of the stage**	**Roles of the quality specialist in each stage**
Deciding to foster new ideas to help the progress of the company	D	
Bringing about the emergence of an idea which may provide a response to a need of society	CR	In most innovation processes, the quality specialist who will assist the project has not yet been designated at this stage of the creative research. – If he/she has already been designated, his/her role is limited to suggesting to the project manager to outline the CQT to be adopted in exchanges with future users of the innovation by providing methodological assistance, if necessary.
Briefly characterizing products or services which could benefit from the innovation	FS	The quality specialist at the stage of the feasibility studies may not yet be assigned to the project. If he/she has been, he/she simply observes the work carried out by the first few members of the project, and suggests to the project

		manager to start to address the design of the CQF associated with previous CQT, and the design of these CQF emitters.
Identifying possible clients for these future products or services	EI	The role of the quality specialist is to suggest to the project manager to take advantage of the identification of future economic parties who are likely to use the innovation to continue to configure these specific components of the CQP. When the project manager gives the marketing department the operational responsibility of these impact studies, the quality specialist assists by trying to elicit information which can provide guidelines in how to prompt these parties' future competitive quality perceptions. The project manager assumes full responsibility for the performance of these CQF components. The role of the quality specialist is limited to making suggestions based on the observations he/she will make.
Characterizing future uses for products or services which would incorporate the innovation	EI	The role of the quality specialist is to guide the research among information on the characterization of the future uses of the innovation, which helps to highlight the observable events of the CQF impacts produced by the innovation on its future users on their competitive quality perceptions. This work is often a coproduction with the marketing department when it exists. The role of the quality specialist is to highlight the information produced by these studies which will be used to design the CQF aimed at project funders and members of the project team.
Creating experimental drafts which represent the idea as accurately as	CR	The role of the quality specialist is to attract the attention of the project manager to the importance of starting to envisage the initial scenarios for the process of designing and releasing the CQF which will be aimed at future end

possible		consumers of the products or services incorporating the innovation, and to lay the foundations, in the construction of drafts, for various competitive quality fuel which will be aimed at these stakeholders. He/she may suggest some simple methodological tools. He/she must also attempt to identify the foundations for the primary information which will then be used to design various CQFs aimed at the bodies financing the project and to the various future members of the project team. He/she must also attempt to identify the basis of the future CQF emitters. He/she must also attempt to identify the foundations for the future emitters of the CQF.
Conducting initial feasibility tests on the future products or services	FS	The quality specialist must urge the project manager to: – continue the design of the CQT to be adopted in relations with future users of the innovation, and the future users of the associated project; – start to envisage some characteristics of the scenarios of the future processes for the production and use of quality in relations with these two classes of stakeholder; – start to identify primary information which could be used to design the CQF of the products or services which the project could offer to these two classes of stakeholder and their emitters; – track these initial ideas.
Deciding to initiate prospective studies on how to use the idea to benefit the company	D	

Conducting prospective studies on possible socioeconomic uses for the innovation, and its impacts on the company and its position	EI	The quality specialist must suggest to the project manager to complete the design of the CQT to be adopted with all the stakeholders in the project, without getting involved in the operation of these processes. However, he/she must propose modifications to other components of the CQP which has already been started in order to follow the guidelines set out in the CQT.
Deciding to guarantee the ability of the company to mobilize the resources necessary, internally and externally, in order to fully realize its innovation process	D	
Identifying future stakeholders in the innovation project	FS	The role of the quality specialist is simply to attract the attention of the project manager to the possibility of making a very early contribution to designing certain parts of the CQP, to assist him/her in researching the information which could be included in these parts and to refine them as appropriate.
Conducting studies on the nature and feasibility of exchanges with future stakeholders in the project necessary for it to be fully realized	FS	During these feasibility studies, which can take place in several stages, the quality specialist must ensure that the methods of exchange analyzed integrate the feasibility of the CQT designed in order to reap the maximum benefits from the quality support which will come from exchanges with these partners.

		He/she must validate the scenarios of the processes for the production and use of quality in the exchanges with these partners (competitive marketing quality). He/she must qualify the specifications for the CQF of products or services which will be offered as compensation to these partners, and the specifications for the emitters. He/she must design and test the feasibility of the procedures: – for monitoring the ability of the CQF emitters; – for dealing with areas of non-compliance to the CQF specifications; – for monitoring the quality perceived by these partners; – for evaluating what these partners will supply to the innovation project.
Deciding contractual issues with stakeholders	D	
Building a processor responsible for carrying out the innovation process and an external communication network	ORG	The quality specialist must: – suggest to the project manager to supplement, or review the CQT for this class of stakeholder; – completing: - the definition of the marketing quality of these various types of stakeholder, and validating them, - the design of the CQF for the products or services which are offered to them, and their associated emitters, - the qualification of these emitters, - the monitoring of the risks of not releasing CQF in compliance with its specifications,

		- the configuration of the treatment system for areas of non-compliance and monitoring quality perceived by these stakeholders regarding the products or services offered to them, - the method of evaluating the compliance of the contribution to the project with the requirements made by the project manager. These missions are only carried out through direct collaboration between the quality specialist and the project manager.
Deciding to progressively design drafts which give a more accurate representation of the idea	D	
Progressively designing these drafts to accurately represent the configuration of the innovation	CR	The quality specialist must: – suggest to the project manager to continue designing the CQT to adopt the use of the products or services incorporating the innovation in exchanges undertaken with stakeholders; – assist the project manager by working with the company's marketing department, if it exists, to ensure that the information to help design the CQT is collected at this stage. If this department does not exist, he/she can provide methodological tools to help certain members of the project team to carry out such actions; – provide feedback to the project manager regarding the correct execution of these actions. Suggest to the project manager to design scenarios of the processes for producing and releasing CQF aimed at these members.

		Carry out these tasks with the assistance of the project manager and the marketing department. He/she may suggest appropriate methodological tools. The quality specialist can ask the project manager to ensure that the specifications of the various CQFs and their emitters can be refined. These are CQFs of that offered to end consumers of the products or services incorporating the innovation, and to all stakeholders in the innovation project. He/she can suggest to the project manager to submit these drafts to representatives of the finance, marketing, purchasing, human resources and other departments responsible for social regulation, which are assigned to the project, in order to give an opinion on the relevance of the CQF and their emitters which are integrated in these drafts. He/she may also, as before, propose specific methodological tools and lead their implementation. He/she may even suggest ideas to change drafts so as to increase their relevance, or the visibility of the fuel and their emitters.
Deciding to carry out prospective studies on the socioeconomic impacts of the future products and services which will incorporate the innovation	D	

Progressively designing drafts which give a representation of future products and services to incorporate the innovation, by describing in particular, the future "interfaces" of the innovation with other components of these products and services	CR	The quality specialist must: – suggest to the project manager to continue designing the CQT to adopt the use of the products or services incorporating the innovation in exchanges undertaken with stakeholders, and which concern certain general characteristics to include in the CQF of the products or services incorporating the innovation; – assist the project manager by working with the company's marketing department, if it exists, to ensure that the information to help create these drafts is collected at this stage. If this department does not exist, he/she can provide methodological tools to help certain members of the project team to carry out such actions; – provide feedback to the project manager regarding the correct execution of these actions. He/she must also: – suggest to the project manager to continue creating competitive marketing quality which will take place in the exchanges with each of these types of stakeholders; – lead the completion of these tasks with the assistance of the project manager and the marketing department; – propose appropriate methodological tools. He must also: – suggest to the project manager to ensure that the specifications of the CQF and their emitters are being refined; – suggest submitting these drafts to representatives of the marketing department, in order to have an opinion on the relevance of the CQF and their emitters which are included in these drafts.

<table>
<tr><td></td><td></td><td>He/she may also propose specific methodological tools and lead their implementation.

He/she may even suggest ideas to change the drafts so as to increase their relevance, or the visibility of this fuel and its emitters.

He/she must finally ensure the design of the CQF integrated into the products or services offered to certain stakeholders targeted in class (P), while complying with provisions in the CQT.

These stakeholders are:

– bodies which finance the design of the production system for products or system incorporating the innovation;

– purchasers of products or services which include the innovation;

– members inside or outside the company who supply the production system with human, material or information-based resources;

– members inside the production system contributing to the production of added value.

He/she continues to propose methodological tools, and ensures that the company's departments which are regularly in contact with these stakeholders are involved in these actions.</td></tr>
<tr><td>Conducting tests on the drafts of future products or services, at various steps in their production, on future clients and on all</td><td>IS</td><td>The quality specialist must attract the attention of the project manager to the possibility of improving the relevance of the CQT to be adopted in exchanges with users and producers of the future products or services, according to the information emerging from these tests.

The role of the quality specialist is to help the marketing department, if it</td></tr>
</table>

future stakeholders in the production system for the products and services		exists, or the project manager, to use these tests to validate the competitive marketing quality of the exchanges with these stakeholders. The quality specialist has full responsibility for the actions to carry out in terms of the characterization of these stakeholders' competitive quality perceptions. To fulfill his/her role, he/she must participate in all the tasks regarding the way these tests are conducted, who conducts them and who analyzes the results. It is his/her role to define how these tasks must take account of the creation of components of the CQP recalled above. He/she requests the project manager for the authorization to activate these decisions at the correct moment during these stages of the impact studies. He/she leads the fulfillment of these tasks, and evaluates the result of them. These evaluations are presented during some of the project reviews. As regards the use of these tests for validating the expected benefits for the company, the quality specialist simply assists the marketing department, which has full responsibility over these actions. The use of tests to give a general characterization of the CQF for the future products or services incorporating the innovation falls entirely within the quality specialist's remit. As regards the quality of the products or services proposed to stakeholders in the production of the products or services incorporating the innovation, the tests help to: – supplementing the design of the CQT;

		– qualifying the emitters of CQF integrated into these products or services; – validating the marketing quality; – starting to envisage methods for supervising the ability of these emitters; – starting to envisage methods for supervising the perceived quality; – verifying the validity of contributions expected from these parties for the production of products or services incorporating the innovation. Under the responsibility of the project manager, the quality specialist must encourage and assist the industrial department of the company to carry out these actions.
Conducting performance tests on future products or services	FS	The quality specialist sets guidelines to ensure that these tests include the actions for the qualification of these CQF emitters aimed at the future users of products or services. Where necessary, he/she participates in carrying out these tests, ensuring that they comply with the preestablished protocol. He/she validates the relevance of the results obtained. He/she encourages the project manager to carry out modifications, where necessary, to: – the CQT to be adopted in exchanges with certain stakeholders on the use of products or services which incorporate the innovation, and on the innovation project; – marketing quality associated with these stakeholders; – specifications for CQF and their emitters. The quality specialist must use the information from these tests to review or supplement the specification of the CQF or their emitters aimed at partners

		contributing to the implementation of the innovation project and at future stakeholders in the BU which will produce these products or services.
Creating a rationale to aid decision-making, examining the probable socioeconomic impacts of implementing the innovation	AD	The quality specialist, at this stage of designing the CQT to be adopted in relations with decision makers, is the "coach" of the project manager. The quality specialist must observe how the project manager naturally undertakes these tasks and suggest supplementary actions to him/her, where necessary, or corrections, at the request of the project manager. The quality specialist must always remain withdrawn in these exchanges with decision makers. The quality specialist is the real director for these role-play scenarios between the members of the project team and the decision makers who will allow the quality of the exchanges contributing to the success of the innovation. However, these scenarios are only submitted to the project manager if he/she so requests. If the project manager does not wish to be assisted in this task, they can be used as guidelines to help the quality specialist fulfill other objectives. In terms of the design of the CQF which must be included in the decision-making assistance process, the quality specialist carries out a technical task of regulation which consists of: – observing how the project manager designs the content and form of the decision-making assistance; – extracting information from these observations which describe the CQF and their emitters; – offering the project manager a pre-analysis of the risks of not prompting the

		competitive quality perceptions desired by decision makers; – completing this analysis, with the project manager, by highlighting any possible risks of decisions likely to reduce the chances of the innovation succeeding, with an unacceptable criticality level; – suggesting a search for solutions to reduce the criticality to acceptable levels; – assisting, or participating in, this quest for solutions, especially taking advantage of feedback as a guide; – estimating the probable effects of the changes undertaken compared with the objectives; – recording the key events of these improvements. Monitoring the aptitude of the project manager to produce this decision-making assistance is a self-monitoring process. This consists of ensuring that the CQF and their emitters are closely compliant with the predefined specifications, just before the production of the decision-making assistance. The quality specialist intervenes at this stage only if the project manager requests him/her to do so. Nevertheless, the quality specialist must monitor how the information released by the decision-making assistance impacts on their competitive quality perceptions. He/she can only do this by interpreting the signals of the decision makers' attitude and behavior. Subsequently, the quality specialist carries out a critical analysis of the deviations between the competitive quality perceptions prompted and those

		which were expected, and the reasons for these deviations. He/she presents this diagnosis to the project manager. Such deviations do not always have negative effects. They can highlight opportunities to capitalize upon in order to increase the chances of the innovation succeeding. This diagnosis can lead to the implementation of corrective action which takes the form of supplementary decision-making assistance.
Deciding to guarantee the ability of the company to adopt a production system for the future products and services	D	
More precisely, identifying stakeholders who will contribute to the development and production of future products and services which will incorporate the innovation	IS	The quality specialist does not have an active role in this stage of the feasibility studies. He/she simply analyzes the decisions which are taken in order to: – help the project manager to supplement the CQT aimed at future stakeholders in the BU which will produce the products or services; – supplement the methods of estimating the quality perceived by this class of stakeholders; – supplement the definition of contributions which will be requested from partners in this class.
Designing drafts of the production and distribution systems for the products and services	ORG	The quality specialist must suggest to the project manager to: – finalize the CQT; – accurately identify the specifications for the compensatory resources requested from them.

		He/she is responsible for monitoring the production of supplementary information for the CQF.
Conducting studies on the ability of the company to design production and distribution systems for the products and services which will incorporate the innovation by taking into account the configurations of the various definitive drafts of the innovation	FS	The quality specialist must: – suggest to the project manager to finalize the CQT to be adopted in exchanges with future producers of the production system for the products or services incorporating the innovation, and the members of the departments within the company responsible for the protection of the environment, people and property; – specify the set of specification for the compensatory resources required from the stakeholders. He/she is responsible for: – overseeing the performance of CQF emitters; – finalizing the design of the methodology: for monitoring quality perceived by these partners of the products or services offered to them; – carrying out the final adjustments to the possible contributions of these partners to the resources and the reputation of the company through the innovation. As before, the quality specialist is responsible for collecting the most relevant information identified in these studies in order to exploit them later on.
Creating a rationale to aid decision-making which details the feasibility of implementing the innovation	AD	The same roles as for the previous assistance to decision-making stage

Examining the steps to take to ensure the intellectual protection of the innovation	ORG	No role for the quality specialist
Deciding to validate the innovation	D	

LEGEND.–

- AD: assistance in decision-making
- D: decision
- CR: creative research
- IS: impact studies
- FS: feasibility studies
- ORG: organization

Bibliography

[ADR 93] ADRIAENSENS B., INGHAM M., VANKERKEM M., *Marketing et qualité totale,* de Boeck, 1993.

[BER 93] BERTALANFFY L., *Théorie générale des systèmes*, Dunod, 1993.

[BOR 01] BORZEIX A., FRAENKEL B., *Langage et travail, communication cognition, action,* CNRS editions, 2001.

[BRA 95] BRAESCH C., HAURAT A., *La modélisation systémique en entreprise*, Hermès, 1995.

[CAG 02] CAGAN J., VOGEL C., *Creating Breakthrough Products*, Prentice Hall, 2002.

[CIA 90] CIALDINI R., *L'influence et la manipulation*, First Editions, Paris, 1990.

[COM 11] COMBE E., MUCCHIELLI J.L., *La compétitivité par la qualité*, Fondapol, 2011.

[CRO 80] CROSBY P., *Qualit: is Free (The Art of Making Quality Certain),* McGraw Hill, New York, 1980.

[DEM 88] DEMING E., *Qualitéy: la révolution du management,* Economica, 1988.

[FER 08] FERNEZ-WALCH S., *Dictionnaire du management de l'innovation*, Vuibert, 2008.

[GAR 12] GARET G., MOCK E., *La fabrique de l'innovation*, Dunod, 2012.

[GOM 96] GOMEZ P.Y., *Le gouvernement de l'entreprise*, InterEditions, Masson, 1996.

[JAC 96] JACOT J.H., MICAELLI J.P., *La performance économique en entreprise*, Hermès, 1996.

[KAP 84] KAPFERER J.N., *Les chemins de la persuasion*, Dunod, 1984.

[KAP 87] KAPFERER J.N., *Les rumeurs*, Seuil, 1987.

[KÉL 92] KÉLADA J., *Comprendre et réaliser la qualité totale*, 2nd ed., Quafec, 1992.

[KÉL 96] KÉLADA J., *Integrating Reengineering with Total Quality*, ASQC Quality Press, 1996.

[LE 94] LE MOIGNE J.L., *La théorie du système général, théorie de la modélisation*, 4th ed., Presse universitaire de France, 1994.

[MAI 11] MAILLARD P., *Les pratiques performantes du TQM (le T scar management)*, Hermès-Lavoisier, 2011.

[MAI 13] MAILLARD P., *Competitive Quality Strategies*, ISTE, London and John & Sons, New York, 2013.

[MIS 99] MISPELBLOM-BEYER F., *Au-delà de la qualité, Alternative sociologique*, 2nd ed., SYROS, 1999.

[MOU 04] MOUGIN Y., *Manager durablement dans l'efficacité*, Afnor, 2004.

[PIL 10] PILLET M., *Améliorer la productivité, déploiement industriel du tolérancement inertiel*, Eyrolles, 2010.

[PRO 88] PROBST G., *Pensée globale et management, résoudre les problèmes complexes*, Les Editions d'Organisation, 1988.

[RIV 06] RIVET P., LÊ P., *Piloter et réussir l'innovation en entreprise*, Maxima, 2006.

[SAI 90] SAISSELIN R.G., *Le bourgeois et le bibelot*, Albin Michel, 1990.

[SHI 97] SHIBA S., GRAHAM A., WALDEN D., *4 Révolutions du management par la qualité totale*, Dunod, 1997.

[THÉ 09] THÉVENET M., *Le bonheur est dans l'équipe*, Eyrolles, 2009.

[TRI 91] TRIPIER P., *Du travail à l'emploi, paradigmes, idéologies et interactions*, Editions de l'Université de Bruxelles, 1991.

[VIN 00] VINCK D., *Pratiques de l'interdisciplinarité, mutations des sciences et de l'enseignement*, Presses universitaires de Grenoble, 2000.

Index

C, D

capital confidence, 206
classes of expectations, 62
competitive quality
  process (CQP), 3, 21, 53, 69, 75
    main functions, 70
    technical functions, 79
  fuel (CQF), 65
    design of, 125, 129, 137, 230
  tactics (CQT), 75
configuration,
  management, 24, 80
creative research, 7, 25, 82, 205, 214, 291, 303
criticality of risks, 110, 163, 170, 199, 202, 207, 251, 253, 261, 265, 295, 314, 317
decision-making
  process, 34, 36, 162
  stages, 9, 10, 36, 69, 279, 306
driver of socioeconomic
  exchanges, 18–21, 78–81, 149, 158, 204, 245
dynamic regulation, 61, 211, 295

E, F, G, H

ensuring benefits, 197, 317
evaluation guidelines, 306
feasibility study, 8, 32–35, 42, 43, 47, 50, 69, 88, 90, 110, 124, 214, 232, 253, 259, 281, 291, 305
  technical stages, 34
functionality, 14, 47
gambles, 77, 185
general systemics, 13
horizontal managerial
  organization, 13–16, 26

I, L, M

impact studies, 7, 29–33, 47, 50, 69, 82, 214, 232–240, 291, 304
innovation, 291, 304
  potential market, 58, 215, 234, 255, 272
  Type I, 39–43, 48
  Type II, 39, 40, 44–49, 51

Type III, 39, 40, 49, 51
Type IV, 39, 40, 52
language, 100, 131, 208, 294
managerial organization
horizontal, 13–16, 26
pyramid, 15, 27, 210
marketing quality, 70–72, 97, 124, 130, 134, 143, 163, 169, 177, 193, 225, 242
competitive, 115, 132, 144, 163, 169–174, 193, 229, 231, 242, 250, 260, 269, 276, 283, 312, 315

O, P, Q

organizational stages, 9, 69, 305
perceived quality, 62–64, 69, 94, 159, 175, 178, 187, 197, 238, 252, 310, 317
primary information, 91, 144, 145, 150, 151, 160–162, 179, 219, 220, 230, 239–242, 252, 257, 264, 283, 284, 314, 316
project
investors, 22, 57, 58, 97, 144, 216, 234, 256, 272
providers, 22, 57, 58, 216, 234, 256, 272
team, 18, 287
pyramid managerial organization, 15, 27, 210
quality
comburant, 134, 135, 160, 178, 180, 185, 187, 204, 220, 281, 311, 316
compliance, 64, 95
department, 205
game, 77

R, S

relational methods, 64, 169
"S.C.A.R" expectations, 62
strategic area of activity, 19–23, 30, 45, 56, 86, 182, 243
support services, 15, 16, 27, 37, 60, 61

Other titles from

ISTE

in

Business, Management and Finance

2014

LESCA Humbert, LESCA Nicolas
Scanning the Business Environment and Detecting Weak Signals

2013

JANSSEN Jacques, MANCA Oronzio, MANCA Raimondo
Applied Diffusion Processes from Engineering to Finance

2012

DAL PONT Jean-Pierre
Process Engineering and Industrial Management

POMEROL Jean-Charles
Decision-making and Action

SZYLAR Christian
UCITS Handbook

2011

LESCA Nicolas
Environmental Scanning and Sustainable Development

LESCA Nicolas, LESCA Humbert
Weak Signals for Strategic Intelligence: Anticipation Tool for Managers

MERCIER-LAURENT Eunika
Innovation Ecosystems

2010

SZYLAR Christian
Risk Management under UCITS III/IV

2009

COHEN Corine
Business Intelligence

ZANINETTI Jean-Marc
Sustainable Development in the USA

2008

CORSI Patrick, DULIEU Mike
The Marketing of Technology Intensive Products and Services

DZEVER Sam, JAUSSAUD Jacques, ANDREOSSO Bernadette
Evolving Corporate Structures and Cultures in Asia / Impact of Globalization

2007

AMMI Chantal
Global Consumer Behavior

2006

BOUGHZALA Imed, ERMINE Jean-Louis
Trends in Enterprise Knowledge Management

CORSI Patrick *et al.*
Innovation Engineering: the Power of Intangible Networks

Lightning Source UK Ltd.
Milton Keynes UK
UKOW06n1111090615
253149UK00001B/2/P

9 781848 218208